RESTRICTED

AN 01-75FJC-1

HANDBOOK
FLIGHT OPERATING INSTRUCTIONS

USAF SERIES
T-33A

NAVY MODEL
TO-2

AIRCRAFT

REVISION NOTICE

LATEST REVISED PAGES SUPERSEDE THE SAME PAGES OF PREVIOUS DATE

Insert revised pages into basic publication.
Destroy superseded pages.

Appendix I of this publication shall not be carried in Aircraft on missions where there is a reasonable chance of its falling into the hands of an unfriendly nation.

PUBLISHED UNDER AUTHORITY OF THE SECRETARY OF THE AIR FORCE AND THE CHIEF OF THE BUREAU OF AERONAUTICS

NOTICE: This document contains information affecting the national defense of the United States within the meaning of the Espionage Laws, Title 18, U.S.C., Sections 793 and 794. Its transmission or the revelation of its contents in any manner to an unauthorized person is prohibited by law.

RESTRICTED

15 MAY, 1950
REVISED 24 APRIL 1951

RESTRICTED
AN 01-75FJC-1

Reproduction of the information or illustrations contained in this publication is not permitted without specific approval of the issuing service. The policy for use of Classified Publications is established for the Air Force in AFR 205-1 and for the Navy in Navy Regulations, Article 1509.

LIST OF REVISED PAGES ISSUED

INSERT LATEST REVISED PAGES. DESTROY SUPERSEDED PAGES.

NOTE: The portion of the text affected by the current revision is indicated by a vertical line in the outer margins of the page.

Page No.	Date of Latest Revision	Page No.	Date of Latest Revision
*1	24 April 1951	45	15 November 1950
*1	24 April 1951	46	15 November 1950
*2	24 April 1951	47	15 November 1950
3	15 November 1950	48	15 November 1950
*4	24 April 1951	49	15 November 1950
*4A	24 April 1951	50	15 November 1950
*6	24 April 1951	51	15 November 1950
*7	24 April 1951	52	15 November 1950
*8	24 April 1951	53	15 November 1950
*9	24 April 1951		
*10	24 April 1951		
*11	24 April 1951		
*12	24 April 1951		
*13	24 April 1951		
*14	24 April 1951		
*14A	24 April 1951		
*18	24 April 1951		
18A	15 November 1950		
*20	24 April 1951		
*20A	24 April 1951		
*20B	24 April 1951		
*21	24 April 1951		
22	15 November 1950		
23	15 November 1950		
*24	24 April 1951		
*24A	24 April 1951		
*24B	24 April 1951		
*25	24 April 1951		
26	15 November 1950		
26A	15 November 1950		
*26B	24 April 1951		
29	15 November 1950		

ISBN #978-1-935327-51-6   1-935327-51-8
©2008 PERISCOPE FILM LLC
ALL RIGHTS RESERVED
WWW.PERISCOPEFILM.COM

* The asterisk indicates pages revised, added or deleted by the current revision.

ADDITIONAL COPIES OF THIS PUBLICATION MAY BE OBTAINED AS FOLLOWS:

USAF ACTIVITIES.—In accordance with Technical Order No. 00-5-2.
NAVY ACTIVITIES.—Submit request to nearest supply point listed below, using form NavAer-140: NAS, Alameda, Calif.; ASD, Orote, Guam; NAS, Jacksonville, Fla.; NAS, Norfolk, Va.; NASD, Oahu; NASD, Philadelphia, Pa.; NAS, San Diego, Calif.; NAS, Seattle, Wash.
For listing of available material and details of distribution see Naval Aeronautics Publications Index NavAer 00-500.

Revised 24 April 1951

RESTRICTED
AN 01-75FJC-1

TABLE OF CONTENTS

SECTION I—DESCRIPTION

Para.		Page
1-1	Airplane	1
1-5	Flight Controls	1
1-22	Power Plant Controls	2
1-34	Water-Alcohol Injection & Fuel Filter De-icing	4
1-37B	Jato Controls	4
1-38	Fuel System	4
1-50	Landing Gear	12
1-56	Hydraulic System	12
1-62	Electrical System	12
1-68	Overheat and Fire Warning Units	13
1-70	Pilots' Seats	13
1-72	Canopy	14
1-74A	Attitude Gyro	14
1-75	Operational Equipment	14

SECTION II—NORMAL OPERATING INSTRUCTIONS

Para.		Page
2-1	Before Entering the Airplane	18
2-8	On Entering the Airplane	19
2-11	Fuel System Management	20
2-13	Starting the Engine	20
2-17	Warm-up and Ground Test	20B
2-21	Taxiing Instructions	21
2-24	Before Take-off	24
2-24A	Jato Technique	24
2-25	Take-off	24A
2-26	Climb	24B
2-30	During Flight	24B
2-40	Stalls	25
2-41	Spins	26
2-58	Permissible Acrobatics	27
2-62	Diving	27
2-68	Night Flying	27
2-70	Approach and Landing	28
2-82	Stopping the Engine	29
2-84	Before Leaving the Airplane	29

SECTION III—EMERGENCY OPERATING INSTRUCTIONS

Para.		Page
3-1	Emergency Exit	30
3-4	Fire	30
3-9	Engine Failure During Take-off	31
3-15	Engine Failure During Flight	31
3-21	Fuel System Emergency Operation	32A
3-23A	Tip Tanks—Fuel System Malfunction	32A
3-24	Electrical Failure	32A
3-27	Emergency Salvo Switch	32A
3-29	Wing Flap Emergency Operation	32A
3-31	Landing Gear Emergency Operation	33
3-33	Landing with Wheels Retracted	33
3-35	Landing in Water (Ditching)	34
3-38	Hydraulic System Emergency Operation	34

SECTION IV—OPERATIONAL EQUIPMENT

Para.		Page
4-1	Ventilating	35
4-4	Heating and Pressurizing	35
4-14	Defrosting	36
4-19	Oxygen System	36
4-23	Armament	37
4-26D	Tow Targets	38
4-27	Lighting	38
4-30	Communications Equipment	38
4-35	De-icing Equipment	38A

APPENDIX I—FLIGHT OPERATING DATA

Para.		Page
A-1	Flight Operation Instruction Charts	39
A-7	Take-off Chart	39
A-11	Use of Flight Operation Instruction Charts	39
A-14	Examples of Use of Charts	40

LIST OF ILLUSTRATIONS

Fig. No.	Title	Page
1-1	The Airplane	ii
1-2	Surface Control Lock	2
1-2A	Engine Fuel System	4B
1-3	Fuel System Schematic	5
1-4	Front Cockpit, Left Hand Side	6
1-5	Rear Cockpit, Left Hand Side	7
1-6	Front Cockpit Instrument Panel	8
1-7	Rear Cockpit Instrument Panel	9
1-8	Front Cockpit, Right Hand Side	10
1-9	Rear Cockpit, Right Hand Side	11
1-10	Seat Controls	13
1-11	External Canopy Controls	14
1-12	General Arrangement	15
1-13	Replenishment Diagram	16
2-1	Courses of Fuel Flow	23

Fig. No.	Title	Page
2-1A	Stall Speed Table	26A
2-1B	Turbulent Air Penetration Speed	26A
2-1C	Load Factor versus Airspeed Diagram	26B
2-2	Spin Pattern	26
2-3	Approach Diagram	29
3-1	Landing Gear Controls	33
4-1	Heating and Pressurization Diagram	35
A-1	Combat Allowance Chart	41
A-2	Engine Operating Limits	42
A-3	Airspeed Correction Chart	42
A-4	Take-off Distances	43
A-5	Landing Distances	44
A-6	Climb and Descent Chart	45
A-7	Flight Operation Instruction Charts	46
A-8	Instrument Markings Diagram	52

Revised 24 April 1951

RESTRICTED
AN 01-75FJC-1

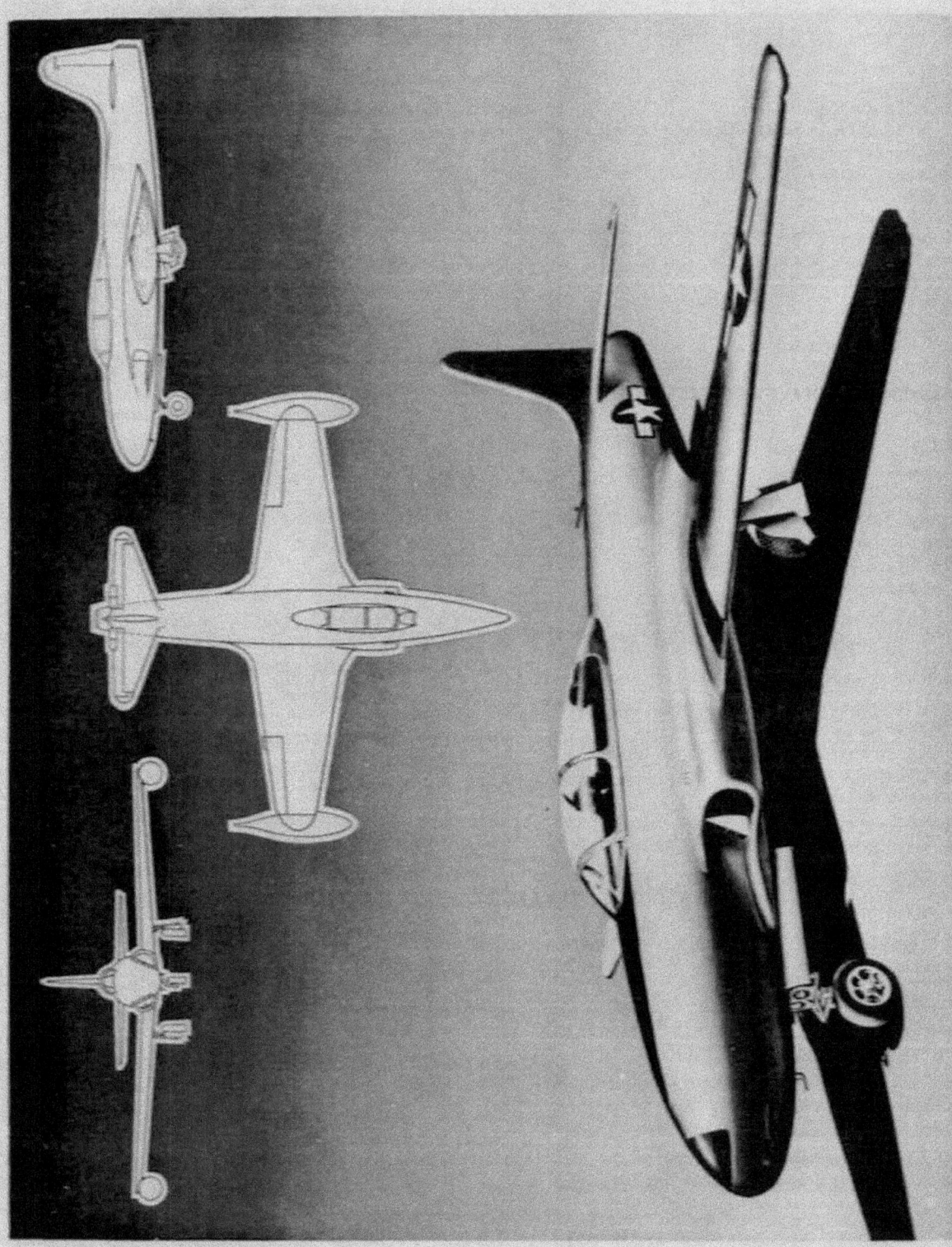

Figure 1-1 — The Airplane

SECTION I

DESCRIPTION

1-1. AIRPLANE.

1-2. TYPE. The T-33A is a two place, tandem, jet propelled, fighter-trainer. Arrangement of the front (student's) cockpit controls is almost identical with the F-80C fighter airplane. Two 50 caliber aircraft machine guns are mounted in the nose.

1-3. DIMENSIONS. The overall dimensions are as follows:

```
Wing span ........................... 38 ft. 10.5 in
Fuselage length ..................... 37 ft.  8 in.
Height (to top of rudder) .......... 11 ft.  8 in.
```

1-4. WEIGHT. The airplane normal gross weight is 11,800 pounds, the maximum gross (full drop tanks, water and ammunition) is 14,250 pounds.

1-5. FLIGHT CONTROLS.

1-6. SURFACE CONTROLS. Conventional stick and rudder pedals, mechanically interconnected, are provided in each cockpit.

1-7. AILERON BOOST. Aileron forces are reduced by a hydraulic aileron booster. The control force reduction is effective after the control stick is moved about two degrees from the neutral position. This system does not destroy the "feel" of the control as a small portion of the total force required (which varies normally with airspeed and rate of application) is supplied by the pilot. The remaining force is supplied by hydraulic pressure.

1-8. AILERON BOOST EMERGENCY SHUT-OFF. (Figures 1-4 and 1-5, references 18A, 20 and 14A, 15.) This switch (or manual lever, on later airplanes) is provided to permit shutting off hydraulic pressure to the booster in an emergency.

1-9. AILERON BOOSTER BY-PASS VALVE. The aileron booster actuating cylinder contains an automatic by-pass valve which opens to permit manual operation if hydraulic pressure fails.

1-10. ELEVATOR. A bungee spring in the elevator control system assists in holding the elevator in the up or down position. This arrangement gives a peculiar feel to the elevator control while on the ground. That is, after the elevator has moved up approximately 20 degrees from the neutral position, it will stay up and, if below that position, will stay down due to the action of the spring. The presence of the spring is not noticeable in flight.

1-11. The spring-loaded elevator tabs located inboard of the trim tabs act to assist the pilot whenever the force on the control stick exceeds approximately 5 pounds.

1-12. RUDDER. The rudder pedals are lightly spring-loaded toward the neutral position to reduce sensitivity.

1-13. TRIM TABS. The trim tabs on the elevator and the left aileron are electrically operated and both are controlled by a single switch on the control stick (figures 1-6 and 1-7, references 41 and 20). Pushing the switch forward or aft operates the elevator tabs, and moving the switch left or right operates the aileron tab.

1-14. A green light (figures 1-6 and 1-7, references 35 and 33) glows when the elevator trim tabs are in the neutral position.

1-15. The rudder tab is not controllable in flight.

1-16. WING FLAPS. The wing flaps are operated by a switch (figures 1-4 and 1-5, references 8 and 16) which controls two electric motors, one for each flap. On later airplanes, the toggle switch is replaced by a lever. The lever operates in the same manner as the switch, except that the levers in the front and rear cockpit are interconnected, to preclude the possibility of having the levers in opposite positions. The position indicator (figures 1-4 and 1-5, references 9 and 10) shows the position of the flaps at all times. On later airplanes the flap position indicator, located on the instrument sub-panel (figures 1-6 and 1-7, references 1A and 40A) is electrically actuated and is calibrated in degrees of flap travel (0—45°). The left and right wing flaps are interconnected so that either motor can operate both flaps if the other motor should fail. There is no emergency mechanical wing flap extension system on this airplane.

1-17. DIVE FLAPS. The dive flaps are controlled by a sliding switch (figures 1-4 and 1-5, references 18 and 14) mounted on the throttle lever which operates an electrically actuated hydraulic valve. The slide is pushed aft to lower the flaps and pushed forward to raise the flaps. The dive flaps cannot be stopped in any intermediate position.

1-18. STUDENT LOCKOUT AND TELL-TALE IN-INDICATOR LIGHT BOX. (Figure 1-6, reference 12.) This control box may be installed in either cockpit and contains seven amber indicator lights and an electrical controls lockout button.

1-19. TELL-TALE INDICATOR LIGHTS. The tell-tale indicator lights show the instructor when the electrical surface controls are being operated by the student. An indicator light is provided to show each of the following operations: elevator tab nose-up and nose-down aileron tab right and left, wing flaps up and down and dive flap down.

1-20. STUDENT LOCKOUT. When the lockout button is pushed both amber lockout indicators on the instrument panels (figures 1-6 and 1-7, references 21 and 17) will glow and the trim tabs, wing flap (Early airplanes only), dive flap and starting fuel sequence controls in the other cockpit will be inoperative. The controls may be restored to operation by pushing the "LOCKOUT RELEASE BUTTON" (figures 1-4 and 1-5, references 3 and 20) in either cockpit.

WARNING

Check the position of the wing and dive flap switches in each cockpit before pushing the "LOCK-OUT RELEASE" button, to avoid unexpected operation of the wing and/or dive flaps.

1-21. SURFACE CONTROL LOCK. (Front Cockpit Only). The surface control lock consists of a tubular bracket attached to the rudder pedals and to the control stick by a thumbscrew (figure 1-2). The lock is clipped to the right-hand side panel when not in use.

1-22. POWER PLANT CONTROLS.

1-23. GENERAL. The engine in this airplane incorporates two separate fuel control systems, (figure 1-3) with a dual engine driven fuel pump. One side of the

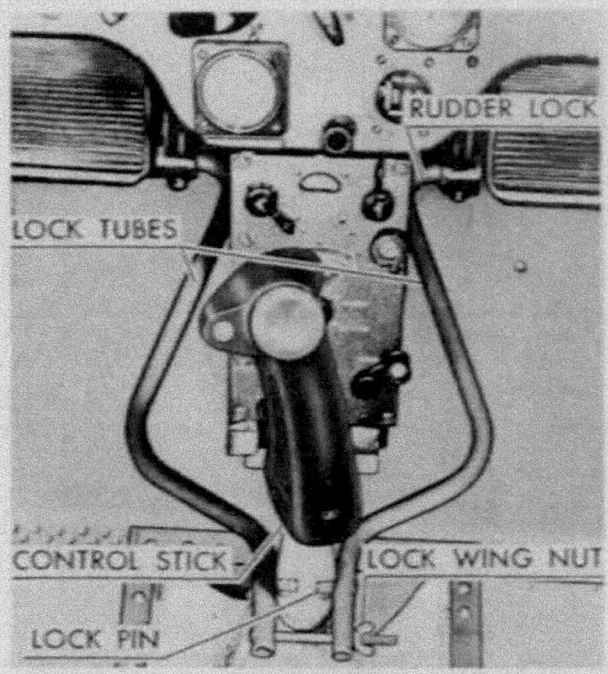

Figure 1-2 — Surface Control Lock
(Front Cockpit Only)

pump supplies the normal fuel system, the other supplies the emergency fuel system. The pump is designed so that in the event one side fails, the other can still supply fuel to the engine. The normal fuel system control is known as the Bendix Gas Turbine Control. The emergency system control is known as the Rochester Control. A pressure switch is installed for automatic actuation of the emergency system when necessary and a starting fuel sequence control is installed for automatic starting.

1-24. OIL SYSTEM. The engine oil system is automatic and requires no controls. A pressure gage is located on each instrument panel (figures 1-6 and 1-7, references 26 and 23). The system is an integral wet sump type with pressure and splash lubrication. The supply reservoir is contained in the accessory drive gear housing. Oil level in the reservoir may be determined by a bayonet gage located on the right side of the accessory gear casing. The reservoir is filled with 12 (U.S.) quarts of oil, Specification AN-O-9, grade 1010.

1-25. THROTTLE. The throttle is the only power control in this airplane. The throttle regulates the fuel pressure to the burner fuel jets on the engine and the resulting fuel flows determine the engine rpm. When the throttle is full aft in the position marked "OFF," it shuts off all fuel to the engine burner ring except the fuel which goes through the starting fuel sequence control. The throttle is connected directly to the Bendix Main Fuel Control which is an all speed governor. The Bendix control attempts to maintain constant engine rpm for any throttle setting, regardless of altitude or airspeed. A maximum throttle position stop is provided, which protects the engine from overspeeding whenever the engine is operating on the Bendix control alone. The Bendix Control also limits acceleration temperature, thereby making it possible to open and close the throttle at any rate, without danger of damaging the engine or incurring flame-out. A throttle linkage is provided on the engine between the Bendix Main Fuel Control and the Rochester Control. The Rochester Emergency Fuel System Control consists of a throttle valve, an altitude compensated relief valve (or Barometric), and a solenoid operated by-pass valve. The relief valve in the emergency fuel control is adjusted to provide 100% engine rpm on a 100°F day. Available full throttle rpm will vary according to free air temperature and altitude. The altitude compensation in the emergency fuel control attempts to maintain constant engine rpm for a given throttle setting, regardless of changes in airplane altitude, however, in flight, overspeeding will generally be possible on the emergency system. The solenoid operated by-pass valve is normally open. Putting the Emergency Fuel switch into the "EMERGENCY" position closes this valve, or if the Emergency Fuel switch is in the "TAKE-OFF & LAND" position, failure of normal fuel system operates a pressure switch to close the by-pass valve which puts the emergency fuel control into operation.

1-26. AUTOMATIC STARTER SWITCH. (Figures 1-8 and 1-9, references 4 and 13.) The starter switch operates automatically in that it does not have to be held in the "START" position but the starter will continue to run until the engine reaches approximately 15% rpm and will then automatically shut off. If it is desired to stop the starter before it automatically cuts off, as in a false start, the switch must be pushed to the "STOP-START" position. The center position is "OFF."

1-27. EMERGENCY FUEL SWITCH. (Figures 1-4 and 1-5, references 23 and 4.) This switch has three positions "OFF," "EMERGENCY," and "TAKE-OFF and LAND." When the switch is set to the "TAKE-OFF and LAND" position, the emergency system will be actuated whenever the pressure on the normal system falls below the pressure switch setting (approximately 45 lbs.). This setting is low enough so that with the throttle in the idle position, the emergency system should not be actuated unless there is a failure in the normal system. Once the emergency system has been actuated, it is necessary to move the switch to the "OFF" position to return control to the normal fuel system. When the emergency switch is set to "EMERGENCY," the emergency fuel control is actuated regardless of the pressure switch.

1-28. EMERGENCY FUEL SYSTEM INDICATOR LIGHTS. (Figures 1-6 and 1-7, references 44 and 39.) There are three indicator lights provided; one red, one green, and one amber. The red light comes on if the landing gear is down and the emergency fuel switch is in the "OFF" position. The green light turns on and the red light turns off when the emergency fuel switch is placed in the "TAKE-OFF and LAND" position. The amber light turns on and the green light continues to stay on, if the emergency fuel switch is in the "TAKE-OFF and LAND" position and the emergency fuel control is in operation. When the emergency fuel switch is placed in the "EMERGENCY" position, the green and amber lights come on and the red light goes out.

1-29. EMERGENCY FUEL CHECKOUT SWITCH. (Figure 1-8, reference 15.) This switch is located on the right-hand shelf near the cabin altimeter. It is provided to check out the emergency fuel system on the ground. When the emergency fuel switch is "OFF" and this switch is actuated, the main fuel pump supply is by-passed and at the same time power is supplied to operate the emergency fuel control provided the pressure switch closes as it should.

1-30. STARTING FUEL SEQUENCE SWITCH. (Figures 1-4 and 1-5, references 10 and 5.) This airplane incorporates an automatic and manual engine starting system. When the switch is placed in the "MANUAL" position, the solenoid valve in the emergency fuel control is closed causing the emergency fuel system to build up pressure. The by-pass valve in the normal side of the main fuel pump remains closed allowing the normal system pressure to build up. When the switch is placed in "AUTO," the same changes take place as in the "MANUAL" position and, in addition, the starting fuel sequence control is energized allowing fuel to go first to the two burners which have ignitor plugs installed and then as the pressure builds up to all other fuel nozzles. The starting sequence may be stopped at any time by placing the switch in the "OFF" position.

1-31. AIR START IGNITION SWITCH. (Figures 1-4 and 1-5, references 11 and 7.) This switch is used to control the ignitor plugs during air starts. The ignition is turned on automatically whenever the starter is operated. In flight ignition is accomplished by operating the air start ignition switch. Since the ignitor plug life is materially shortened by operation of the ignition, a time delay switch is incorporated to automatically limit the duration of the ignition to 45 to 60 seconds. When

Section I
Paragraphs 1-32 to 1-40

the air start ignition switch is pushed to "START" and released, the ignition will continue for the duration of the time delay or until the "NORMAL-OFF" ignition switch (figure 1-8, reference 3) is turned to "OFF."

1-32. IGNITION "NORMAL-OFF" SWITCH. (Figure 1-8, reference 3.) This switch is provided for ground testing only and must be kept in the "NORMAL" position at all other times.

1-33. IGNITION TEST SWITCH. (Figure 1-8, reference 8.) The ignition test switch has been removed from the airplanes.

1-34. WATER-ALCOHOL INJECTION AND FUEL FILTER DE-ICING.

1-35. FLUID INJECTION SYSTEM. The system is independent and consists of two tanks (located in the air intake ducts), with a total capacity of 50 U.S. gallons, an electrically driven pump, a combination filter and shut-off valve, a pressure transmitter and a dual ring of spray nozzles. Also included is an actuating cylinder which automatically shuts off pressurizing air to the cockpit while the fluid injection system is operating. This is to prevent noxious fumes from entering the cockpit.

1-36. Use of fluid injection will give increased thrust for short periods and is especially useful for short field take-offs or emergencies in warm weather. The use of fluid injection is prohibited at ground air temperatures of below +32°F, and above 10,000 feet.

1-37. FLUID INJECTION SWITCH. The fluid injection switch on the left-hand shelf (figures 1-4 and 1-5, references 22 and 17) turns on the electrically driven fluid injection pump, provided the throttle is almost fully advanced. This is true because the throttle is linked to another switch in series with the fluid injection switch which prevents a completed circuit if the throttle is not near its full open position. This second switch is automatically operated and is provided primarily to help prevent flame-outs and engine damage caused by injecting fluid at low engine rpm. It is still possible, however, to cause flame-out by turning on fluid injection before the engine has accelerated to above 90% rpm. This is true because the throttle can be advanced at a rate faster than the engine can follow.

WARNING

Never turn fluid injection on below 90% rpm, or above 10,000 feet altitude due to the possibility of flame-out and engine damage.

1-37A. FUEL FILTER DE-ICING. Provisions for alcohol de-icing of the low pressure fuel filter are included in later T-33A airplanes. For information on fuel filter de-icing see paragraph 4-35, Section IV.

1-37B. JATO CONTROLS.

1-37C. Jato firing is controlled electrically by a "JATO-GUNS" arming switch in the front cockpit (figure 1-6, reference 29), a master switch in the rear cockpit (figure 1-5, reference 24) and the gun trigger switches on both control sticks. When the "JATO-GUNS" transfer switch is placed in the "JATO" position and the master switch in the rear cockpit is closed, the indicator above the switch glows and the JATO units may be fired by pressing the gun trigger switch in either cockpit. After the units are fired, the "JATO-GUNS" switch must be returned to the "GUNS" position to restore the function of the gun trigger switch. The jato units are jettisoned by pulling the jettison handle in the front cockpit (figure 1-4, reference 38).

1-37D. On later airplanes, serial numbers 50-402 and subsequent, JATO firing is controlled through the arming switch (29, figure 1-6) on the front instrument sub-panel and a firing button on the left shelf (12A, figure 1-4; and 11A, figure 1-5) in each cockpit. When the arming switch is placed in the "ARM" position, the "JATO READY" warning light on both instrument panels glow, indicating that the circuit is ready for firing. After the circuit is armed, pressing the JATO firing button will fire the JATO units.

1-38. FUEL SYSTEM.

1-39. GENERAL. Fuel is carried in four groups of tanks, as shown on figure 1-3. The drop tanks are carried on bomb shackles at the wing tips. Under normal operating conditions, all fuel is transferred to the fuselage tank before being fed to the engine. Fuel transfer is automatically controlled by three float valves within the fuselage tank. The leading edge tank float valve and the wing tank float valve are located one and two inches respectively below the drop tank float valve. When the fuel level of the fuselage tank is above any of the float valves, the respective valve will close. The fuselage tank fuel level is maintained at each float valve level until the corresponding group of tanks is empty.

1-40. Under emergency operating conditions, fuel from the leading edge and wing tanks (but not the drop tanks) may be made to by-pass the fuselage tank. This by-pass condition is controlled by a switch (figures 1-4 and 1-5, references 24 and 19). Failure of the electrical power supply during by-pass operation will cause the system to return to normal operation.

1-40A. **FUEL SPECIFICATIONS AND GRADES.** Fuels used with J33-A-35 engines and modified airplanes with J33-A-23 engines must conform to:

SPECIFICATION MIL-F-5624 (AN-F-58), GRADE JP-3—For all normal operation, including starting.

SPECIFICATION MIL-F-5672 (AN-F-48), GRADE 100/130 (GASOLINE). As an alternate fuel.

1-40B. Fuels used with unmodified airplanes J33-A-23 engines must conform to:

SPECIFICATION MIL-F-5572 (AN-F-48), GRADE 100/130.—In the leading edge fuel tanks (for starting and during purging when stopping the engine).

SPECIFICATION MIL-F-5616 (AN-F-32), GRADE JP-1.—In the remaining tanks (for engine operation).

FUEL QUANTITY DATA
(U.S. GALLONS)

Tanks	No. Tanks	Usable Fuel (Ea.)	Usable Expansion Space (Ea.)	Unusable Fuel (Ea.) Level Flight	Total Volume (Ea.)
Drop Tanks	2	165	0*	0.5	165.5
Leading Edge	2	52	0*	**	**
Wing	2	77	0*	**	**
Fuselage	1	95	0*	**	**
Total Airplane	7	683	0*	12.6	695.6

* All tanks have the usual expansion space however this is not available for "stuffing" purposes, since fuel in this space drains overboard.

** Not determined for separate tank groups.

1-41. **FUEL TANK SELECTOR SWITCHES.**
(Front Cockpit Only.)

1-42. **DROP TANKS.** The drop tank selector switch (figure 1-4, reference 27) operates a solenoid valve which admits air pressure from the engine compressor into the drop tanks. This air pressure forces fuel from the drop tanks into the fuselage tank when the drop tank float valve is open. In case of electrical failure the solenoid valve will automatically open and fuel will be fed from the drop tanks to the fuselage tank.

1-43. **LEADING EDGE TANKS.** The leading edge tank switch (figure 1-4, reference 26) turns on a booster pump in each leading edge tank. These pumps transfer fuel into the fuselage tank when the leading edge tank float valve is open.

1-44. **WING TANKS.** The wing tank switch (figure 1-4, reference 25) turns on a transfer pump in each wing tank. These pumps force fuel into the fuselage tank when the wing tank float valve is open.

1-45. **FUSELAGE TANK.** The fuselage tank and bypass switch located in the front cockpit (figure 1-4, reference 24) has three positions. In the upward "FUS" position of the switch, the fuselage tank booster pump is turned on to supply fuel under pressure to the engine driven fuel pump. In the downward "BYPASS" position of the switch, the fuselage tank booster pump is shut off and the electrically operated bypass valves are reset, causing fuel in the wing tanks and leading edge tanks to bypass the fuselage tank. In the center "OFF" position of the switch, the fuselage tank bypass valves are set for normal operation but the fuselage tank booster pump is off. A guarded master switch located in the rear cockpit (figure 1-5, reference 19) has two positions "NORMAL" and "BY-PASS." When the master switch is in the "NORMAL" position, the switch in the front cockpit controls the circuit. When the master switch is in the "BY-PASS" position the fuel system is in the by-pass condition and the front cockpit has no control of the circuit.

1-45A. **MAIN FUEL VALVE SHUT-OFF SWITCH.** (Figures 1-4, and 1-5, reference 13A.) On later airplanes the ground fuel shut-off valve (figure 1-3) may be operated in an emergency from either cockpit to shut off all fuel to the engine section.

1-46. **DROP TANK RELEASE CONTROLS.** The drop tanks are carried on the wing tip bomb shackles and are normally released by placing the bomb selector switch (figures 1-4 and 1-5, references 32 and 22) in the "ALL" or "TRAIN" position and pushing the bomb and drop tank release button on the control stick. With the bomb selector switch in the "ALL" position, pushing the bomb and drop tank release button will release both tanks simultaneously. When the bomb switch is in the "TRAIN" position, pushing bomb and drop tank release button will release the left tank first, and pushing it again will release the right tank. The tanks may also be released by pushing the emergency bomb salvo switch (figures 1-6 and 1-7, references 40

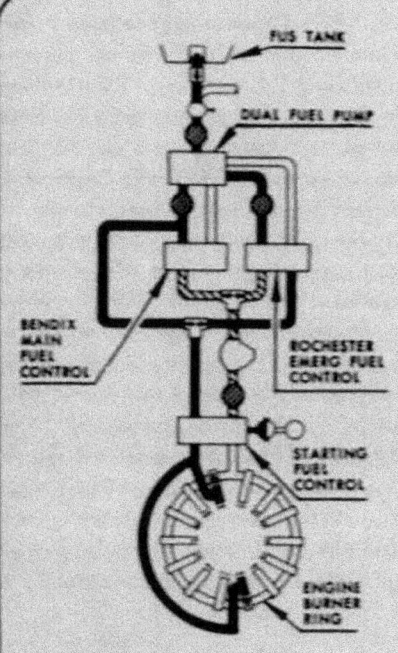

STARTING FUEL FLOW (AUTOMATIC)

Fuel is directed to burners number 7 and 14 until ignited and then to all burners automatically by the Starting Fuel Control.

NORMAL FUEL FLOW

Main fuel pump pressure is directed through the Bendix Speed Density Control to main burner ring. Emergency pump pressure is bypassed.

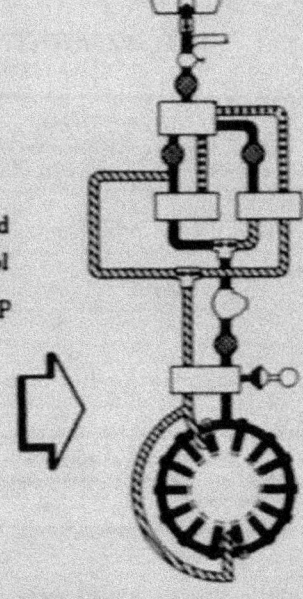

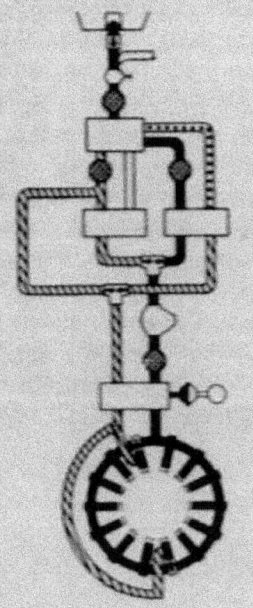

EMERGENCY FUEL FLOW

Emergency fuel pump pressure is directed through Rochester Emergency System Control to main burner ring.

WARNING

When operating on emergency fuel system use extreme care to avoid engine overspeeding.

■ FUEL FLOW 〰〰 STATIC FUEL ⋯⋯ BYPASS FUEL

Figure 1-2A — Engine Fuel System

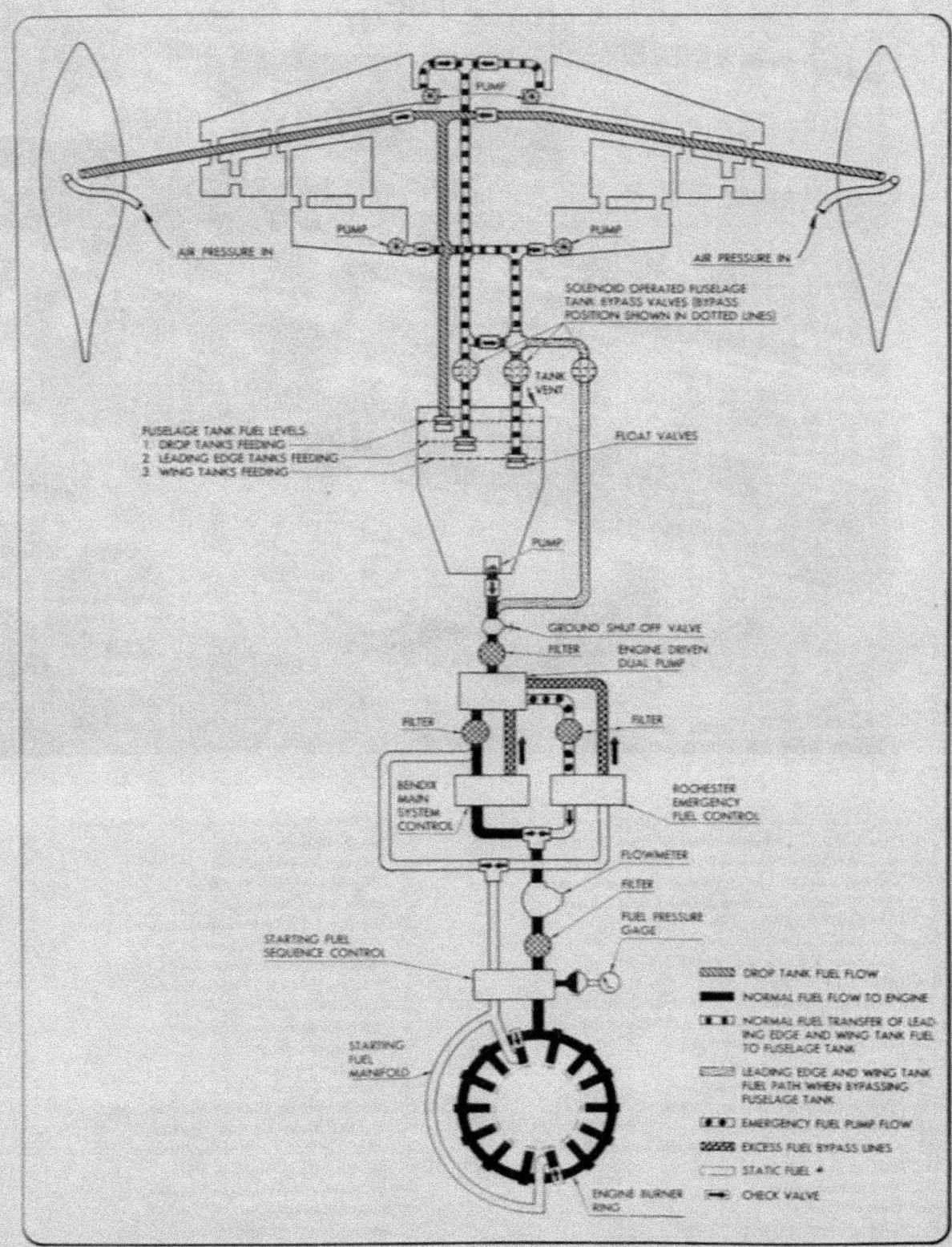

Figure 1-3 — Fuel System Schematic

Section I

RESTRICTED
AN 01-75FJC-1

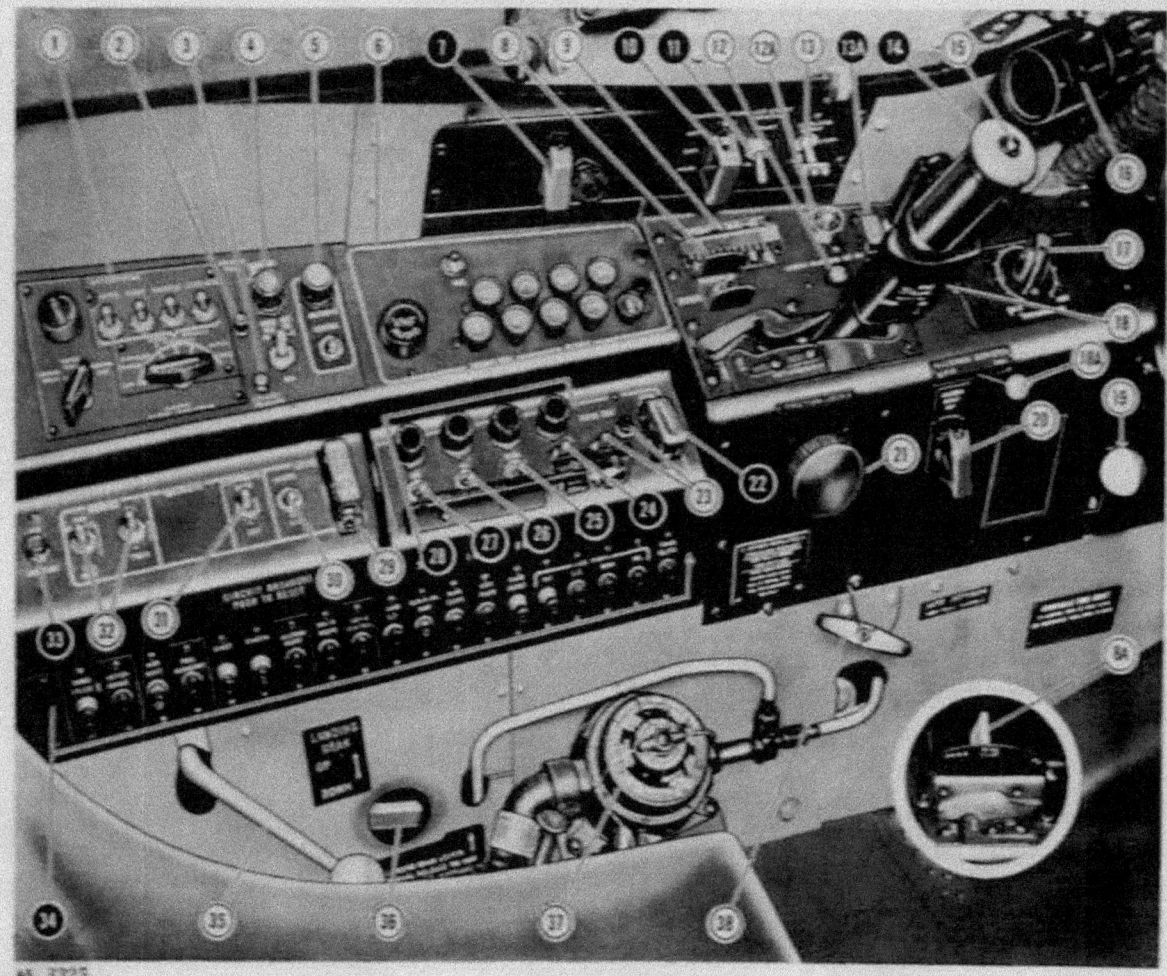

1. Interphone Control Panel (Later Airplanes)
2. Circuit Breaker—Marker Beacon Receiver
3. Student Lockout Release
4. Interphone—AN/ARC-3 Selector (Early Airplanes)
5. AN/ARC-3 Radio Control Switch and Indicator (Early Airplanes)
6. AN/ARC-3 Radio Controls
7. Fuel Filter De-icer and Warning Light
8. Wing Flap Switch
8A. Wing Flap Control Lever (Later Airplanes)
9. Wing Flap Position Indicator (Early Airplanes)
10. Starting Fuel Sequence Switch
11. Air Start Ignition Switch
12. Landing Gear Horn Cutout
12A. Jato Firing Button (Later Airplanes)
13. Landing-Taxi Light Switch
13A. Main Fuel Valve Shut-off Switch (Later Airplanes)
14. Throttle
15. Microphone Button
16. Fluorescent Light
17. Cockpit Heat Control
18. Dive Flap Switch

18A. Aileron Boost Shut-off (Later Airplanes)
19. Hydrofuse Reset Knob
20. Aileron Boost Shut-off
21. Throttle Friction Control
22. Fluid Injection Switch
23. Emergency Fuel Switch and Circuit Breaker
24. Fuselage Tank and By-pass Switch
25. Wing Fuel Tanks Switch
26. Leading Edge Fuel Tank Switch
27. Drop Tanks Fuel Switch
28. Fuel Tank Indicator Lights
29. Guns-Camera Switch
30. Gun Heater Switch
31. Chemical Tanks Switch
32. Bomb Arming and Selector Switches
33. Air Start Ignition Circuit Breaker
34. Left Hand Circuit Breaker Panel
35. Landing Gear Control Lever
36. Landing Gear Downlock Release
37. Oxygen Regulator
38. Jato Jettison Control

● Indicates power plant and fuel system controls and instruments.

Figure 1-4 — Front Cockpit, Left Hand Side

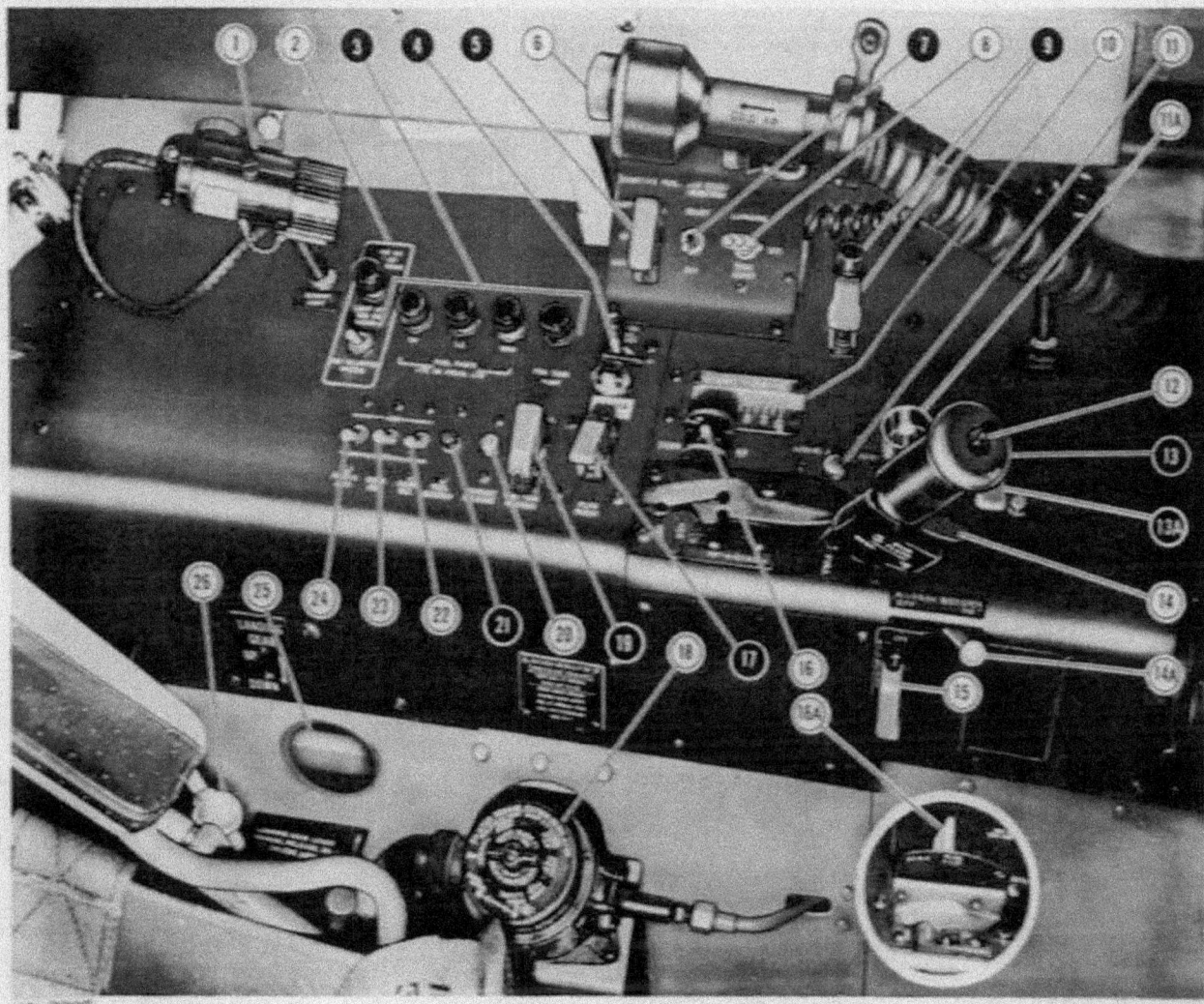

1. Cockpit Light
2. Interphone AN/ARC-3 Radio Switch and Indicator Light (Early Airplanes)
3. Fuel Tank Indicator Lights
4. Emergency Fuel Switch
5. Starting Fuel Sequence Switch
6. Cockpit Ventilator
7. Air Start Ignition Switch
8. Landing-Taxi Light Switch
9. Fuel Filter De-icer and Warning Light
10. Wing Flap Position Indicator (Early Airplanes)
11. Landing Gear Horn Cutout
11A. Jato Firing Button (Later Airplanes)
12. Microphone Button
13. Throttle
13A. Main Fuel Valve Shut-off Switch (Later Airplanes)
14. Dive Flap Switch
14A. Aileron Boost Shut-off (Later Airplanes)
15. Aileron Boost Shut-off Switch
16. Wing Flap Switch
16A. Wing Flap Control Lever (Later Airplanes)
17. Fluid Injection Switch
18. Oxygen Pressure Regulator
19. Fuselage Tank and By-pass Master Switch
20. Student Lockout Release
21. Fuel By-pass Circuit Breaker
22. Bomb Selector Master Switch
23. Bomb Arming Master Switch
24. Jato-Guns Master Switch
25. Landing Gear Downlock Release
26. Landing Gear Control Lever

Indicates power plant and fuel system controls and instruments.

Figure 1-5 — Rear Cockpit, Left Hand Side

Section I

RESTRICTED
AN 01-75FJC-1

1. Accelerometer (Moved to right side on later Airplanes)
1A. Wing Flap Position Indicator (Later Airplanes)
2. Oxygen Pressure Gage
3. Oxygen Blinker
4. Altimeter
5. Standby Magnetic Compass
6. Turn and Bank Indicator
7. Fluid Injection Pressure Gage
8. Landing Gear Position Indicator Lights
9. Airspeed Indicator
10. Attitude Indicator
11. Rate of Climb Indicator
12. Student Lockout and Tell-Tale Indicator Light Box
13. Tailpipe Temperature Indicator
14. Tachometer
15. Gyrosyn Compass Master Indicator
16. Fuel Pressure Gage
17. Gyro Instrument Warning Light
18. Turn and Bank Warning Light
19. Canopy Latch Warning Light
20. Radio Compass Indicator
21. Student Lockout Warning Light
22. Canopy Locking Handle
23. Fire Warning Light and Circuit Test Switch
24. Overheat Warning Light
25. Ammeter
26. Oil Pressure Gage
27. Canopy Jettison Lever
28. Cabin Pressurization Grill
29. Jato Arming Switch and Indicator Light
30. Fuselage Tank Low Level Indicator
31. Fuselage Tank Quantity Gage
32. Parking Brake
33. Fuel Counter
34. Clock
35. Elevator Tab Neutral Indicator
36. Hydraulic Pressure Gage
37. AN/ARN-6 Control Panel
38. Cabin Pressure Dump Valve Lever
39. Gyro Selector
40. Bombs or Tanks Salvo Switch
41. Aileron and Elevator Tab Switch
42. Bombs or Tanks Release Button
43. Gunsight Light Rheostat
44. Emergency Fuel System Indicator Lights

● Indicates power plant and fuel system controls and instruments.

Figure 1-6 — Front Cockpit Instrument Panel

1. Standby Magnetic Compass
2. Fluid Injection Pressure Gage
3. Landing Gear Position Indicator Lights
4. Turn and Bank Indicator
5. Altimeter
6. Attitude Gyro Indicator
7. Airspeed Indicator
8. Rate of Climb Indicator
9. Check List
10. Tailpipe Temperature Indicator
10A. Marker Beacon Indicator
11. Gyrosyn Compass Repeater Indicator
12. Tachometer
13. Gyro Warning Light
14. Turn and Bank Warning Light
15. Canopy Latch Warning Light
16. Radio Compass Indicator
17. Student Lockout Warning Light
18. Canopy Locking Handle
19. Fuel Pressure Gage
20. Aileron and Elevator Tab Control
21. Fire Warning Light
22. Ammeter
23. Oil Pressure Gage
24. Canopy Jettison Lever
25. Overheat Warning Light
26. Lockout Box Receptacle
27. Jato Arming Indicator
28. Bombs or Tanks Release Button
29. Fuselage Tank Low Level Indicator
30. Fuselage Tank Quantity Gage
31. Fuel Counter
32. AN/ARN-6 Control Panel
33. Elevator Tab Neutral Indicator
34. Cabin Pressurization Grill
35. Parking Brake
36. Hydraulic Pressure Gage
37. Clock
38. Bomb or Tanks Salvo Switch
39. Emergency Fuel System Indicator Lights
40. Accelerometer (Moved to right side on later Airplanes)
40A. Wing Flap Position Indicator (Later Airplanes)
41. Oxygen Cylinder Pressure
42. Oxygen Blinker

● Indicates power plant and fuel system controls and instruments.

Figure 1-7 — Rear Cockpit Instrument Panel

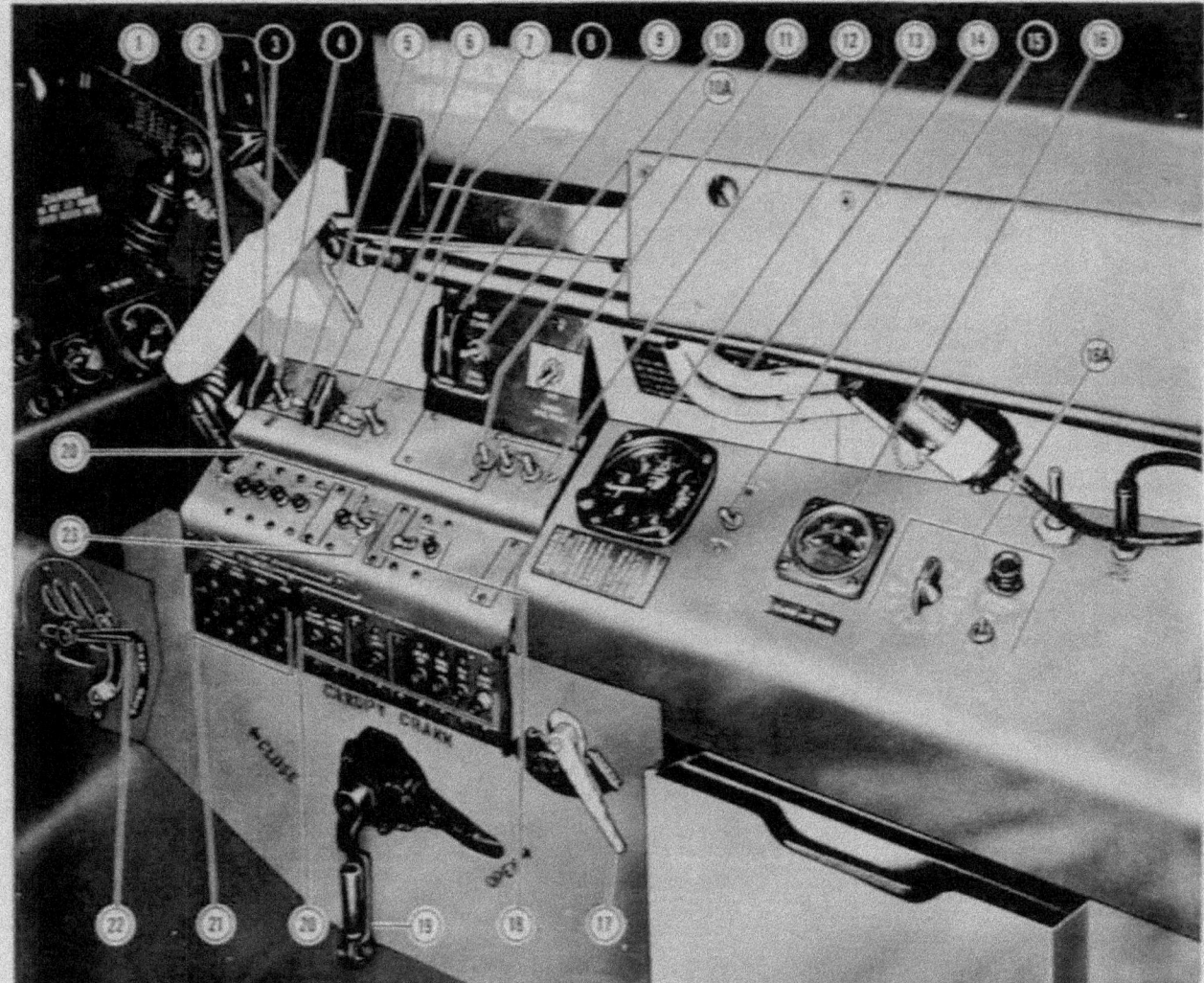

1. Canopy Locking Handle
2. Canopy Jettison Lever
3. Ignition Normal-Off Switch
4. Starter Switch
5. Battery Switch
6. Generator Switch
7. Pitot Heater Switch
8. Ignition Test Switch (Deleted)
9. Canopy "OPEN" Button
10. Canopy "CLOSE" Switch
10A. Fuselage Lights Switch
11. Emergency Hydraulic Pump
12. Navigation Lights Switches
13. Cabin Altimeter
14. Surface Controls Lock (Stowed)
15. Emergency Fuel Checkout Switch
16. Cabin Air Temperature Gage
16A. Code Selector and Signal Light Switch
17. Emergency Hydraulic System Selector
18. Auxiliary Windshield Defroster Switch and Circuit Breaker
19. Canopy Manual Handcrank
20. Right Hand Circuit Breakers
21. Flight Instrument and Spare Fuses
22. Cabin Pressurization Grill
23. Inverter Test Switch

● Indicates power plant and fuel system controls and instruments.

Figure 1-8 — Front Cockpit, Right Hand Side

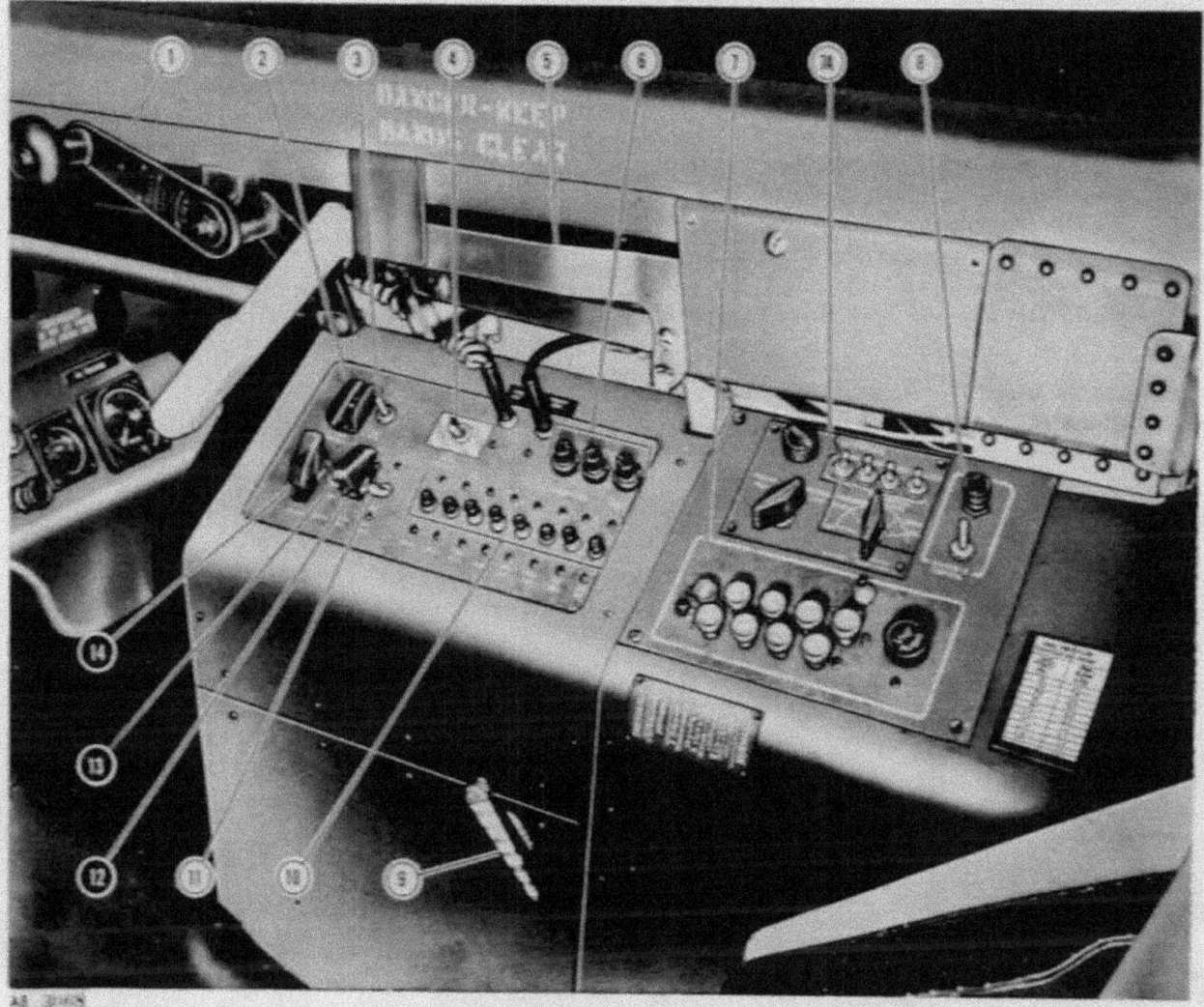

1. Canopy Locking Handle
2. Canopy "OPEN" Button
3. Canopy "CLOSE" Switch
4. Emergency Hydraulic Pump Switch
5. Canopy Jettison Lever
6. Gyro Instrument and Spare Fuses
7. AN/ARC-3 Radio Control Panel
7A. Interphone Control Panel (Later Airplanes)
8. AN/ARC-3 Radio Control Switch and Indicator Light (Early Airplanes)
9. Emergency Hydraulic System Selector
10. Right Hand Circuit Breakers
11. Auxiliary Defroster Switch
12. Battery and Generator Switch
13. Starter Switch
14. Ignition Normal-Off Switch

● Indicates power plant and fuel system controls and instruments.

Figure 1-9 — Rear Cockpit, Right Hand Side

Section I
Paragraphs 1-47 to 1-63

and 38) which permit dropping the bombs or tanks simultaneously in an emergency without presetting the bomb selector switches.

1-47. FUEL TANK INDICATOR LIGHTS. An indicator light for each group of tanks is located on the left-hand shelf (figures 1-4 and 1-5, references 28 and 3). The drop tank, leading edge tanks and the wing tanks indicator lights glow whenever the respective switches are "ON" and the fuel pressure in the lines is below 5 psi. The fuselage tank indicator light glows whenever the fuselage tank booster pump is in operation.

1-48. FUEL QUANTITY INDICATORS. The fuel gage (figures 1-6 and 1-7, references 31 and 30) indicates the quantity of fuel in the fuselage tank only. A low level warning light (figures 1-6 and 1-7, references 30 and 29) glows when the quantity of fuel remaining in the fuselage tank drops below 85 U.S. gallons.

1-49. FUEL QUANTITY COUNTER. The fuel quantity counter operates from a flowmeter in the main fuel line to the engine (see figure 1-3). The counter dial must be set to read the amount of usable fuel in the airplane each time the tanks are filled. The reading on the counter dial is in gallons of fuel remaining in the airplane and will be accurate except under the following conditions:

a. Any fuel which is released when the drop tanks are jettisoned will not, of course, be registered on the counter.

b. Any fuel leaking from the tank vents or a fuel line upstream of the flowmeter will not be counted.

c. Any fuel going through the automatic starting control will not be counted, this usually amounts to about five gallons for an automatic ground start.

1-50. LANDING GEAR.

1-51. LANDING GEAR CONTROLS. (The landing gear is controlled by a lever (figures 1-4 and 1-5, references 35 and 26) and is actuated by normal or emergency hydraulic pressure. While the airplane is resting on the landing gear, an automatic down lock device prevents moving the lever out of the "DOWN" position. This down lock can be disengaged in an emergency by simultaneously pushing down on the release control (figures 1-4 and 1-5, references 36 and 25) and moving the lever to the "UP" position.

1-52. The main and nose gear mechanisms are equipped with up locks and down locks. Operation of these locks is automatic.

1-53. A "Stiff-Knee" is provided for installation on the spring cartridge located between the parallel drag struts for each leg of the landing gear to prevent accidental retraction when the airplane is on the ground.

1-54. Four lights (figures 1-6 and 1-7, references 8 and 3) and a warning horn indicate the landing gear position. A green light for each individual gear glows when its respective gear is down and locked. One red light glows and the warning horn sounds if the throttle is closed while any one of the gears is not down and locked. The horn may be silenced by pushing the switch on the left-hand shelf. The switch is automatically reset when the throttle is opened.

1-55. BRAKE CONTROLS. The brakes are operated by conventional toe pedals. A parking brake handle (figures 1-6 and 1-7, references 32 and 35) locks the brakes for extended periods. No emergency brakes are provided.

1-56. HYDRAULIC SYSTEM.

1-57. NORMAL HYDRAULIC SYSTEM. The engine driven hydraulic pump supplies power for operation of the aileron booster, landing gear, dive flaps and cartridge case ejection door.

1-58. NORMAL SYSTEM CONTROLS. A hydrofuse automatically shuts off hydraulic power to the landing gear, dive flaps and case ejection door systems in event of a leak in either system, but does not shut off power to the aileron booster. A knob (figure 1-4, reference 19) permits manual resetting of the fuse to override it in an emergency.

1-59. The cartridge case ejection door is automatically operated by the gun trigger switch.

1-60. EMERGENCY HYDRAULIC SYSTEM. The emergency hydraulic system consists of a separate reservoir, an electric pump and a separate set of lines. The system can be used for only one complete extension of the landing gear.

1-61. EMERGENCY SYSTEM CONTROLS. The emergency hydraulic selector valve lever (figures 1-8 and 1-9, references 17 and 9) opens and closes the line between the electric pump and the landing gear actuating cylinders. The landing gear control lever (figures 1-4 and 1-5, references 35 and 26) must be used in conjunction with the emergency selector valve lever to permit fluid trapped in the actuating cylinders to return to the normal system reservoir. The emergency pump switch is located on the right-hand side of each cockpit (figures 1-8 and 1-9, references 11 and 4).

1-62. ELECTRICAL SYSTEM.

1-63. GENERAL. The 28 volt single-wire electrical system is powered by a 300-ampere generator. In case of generator failure the high capacity battery can be used for limited operation of the fuel transfer pumps.

1-64. The ammeter (figures 1-6 and 1-7, references 25 and 22) is located on the instrument subpanel. All circuits are protected by circuit breakers except the flight instrument circuits which contain fuses (figures 1-8 and 1-9, references 21 and 6). (Spare fuses are located immediately adjacent.) The generator field, case ejection and gun firing circuit breakers are not accessible in flight and some circuits carrying heavier current loads are protected by additional circuit breakers which are also not accessible in flight.

1-65. The system is in operation when the battery switch (figures 1-8 and 1-9, references 5 and 12) and the generator switches are in the "ON" position.

1-66. INVERTERS. Two inverters are provided. If one fails, the other cuts in automatically. The inverters supply current to the gyro instruments and operate whenever the battery switch is "ON." The inverter test switch (figure 1-8, reference 23) is located on the right-hand shelf of the front cockpit.

1-67. EXTERNAL POWER SUPPLY RECEPTACLE. The external power supply receptacle is located in the aft end of the right-hand wing fillet.

1-68. OVERHEAT AND FIRE WARNING UNITS.

1-69. OVERHEAT AND FIRE WARNING LIGHTS. (Figure 1-6, reference 23.) An overheat warning light for the tail section, a fire warning light for the plenum chamber and a test switch for checking the circuit, are located on the instrument sub-panel. The overheat light may be operated by any one of several thermal switches in the tailpipe section. Operation of the light is an overtemperature warning which may indicate a fire or may be caused by exhaust leakage at the tailpipe clamp, improper adjustment of the thermal switch or a short circuit in the circuit. The fire warning light is operated by thermal switches in the plenum chamber and any overtemperature in this compartment would probably indicate a fire.

1-70. PILOTS' SEATS.

1-71. The pilots' seats (figure 1-10) are the conventional non-jettisonable bucket-seat type. The headrest is omitted from the front seat to improve rear cockpit visibility. Both seats are provided with an inertia reel type shoulder harness.

1-71A. SHOULDER HARNESS LOCK CONTROL. A two position (locked-unlocked) shoulder harness inertia reel lock control is located on the left side of the pilots' seats. A latch is provided for positively retaining the control handle at either position of the quadrant. By pressing down on the top of the control handle, the latch is released and the control handle may then be moved freely from one position to another. When the control is in the unlocked position, the reel harness cable will extend to allow the pilot to lean forward in the cockpit; however, the reel harness cable will automatically lock when an impact force of 2 to 3 g's is encountered. When the reel is locked in this manner, it will remain locked until the control handle is moved to the locked and then returned to the unlocked position. When the control is in the locked position, the reel harness cable is manually locked so that the pilot is prevented from bending forward. The locked position

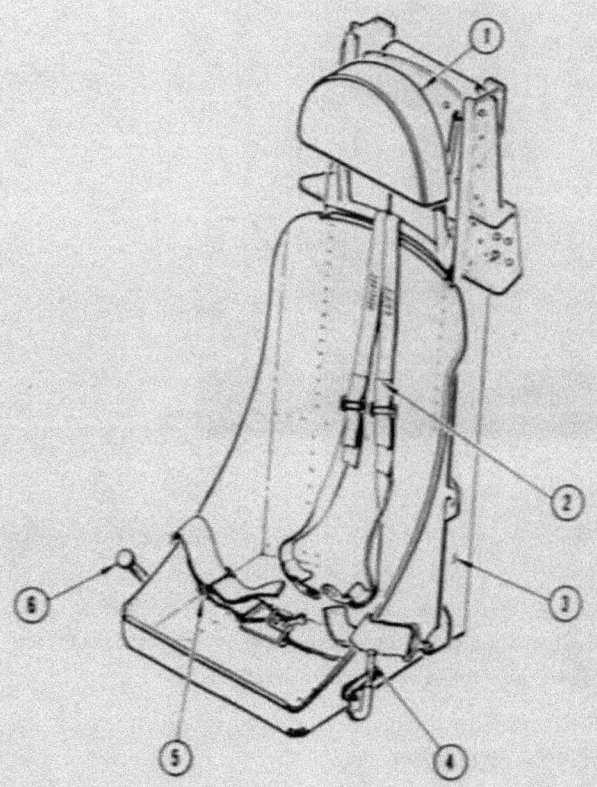

1. HEAD REST (REAR SEAT ONLY)
2. SHOULDER HARNESS
3. SEAT TRACK
4. SHOULDER HARNESS LOCK RELEASE LEVER
5. SAFETY BELT
6. SEAT HEIGHT ADJUSTMENT LEVER

Figure 1-10 — Seat Controls

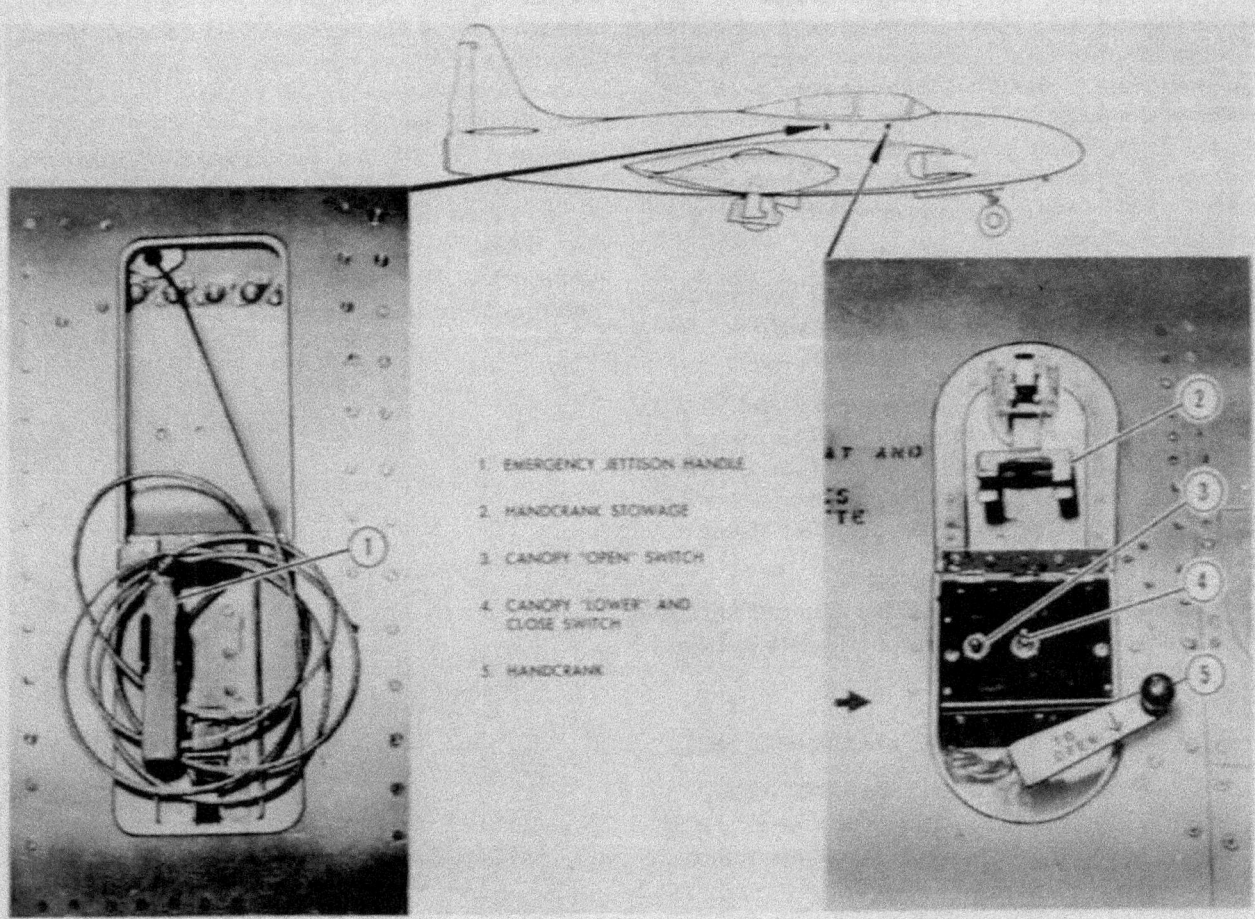

Figure 1-11 — External Canopy Controls

is used only when a crash landing is anticipated. This position provides an added safety precaution over and above that of the automatic safety lock.

1–72. CANOPY.

1–73. The canopy is hinged at the aft end and raised or lowered by an electrically actuated jackscrew, located between the two cockpits. Control switches for normal operation (figures 1-8 and 1-9, references 9 and 2), and a jettison lever (figures 1-8 and 1-9, references 2 and 5) for emergency operation are provided in each cockpit. In addition, a handcrank is provided in the front cockpit for manual operation. The outside control switches, handcrank and external jettison cable are located in wells in the fuselage skin (figure 1-11). The canopy is jettisoned by an explosive charge.

1–74. CANOPY CONTROLS. Two momentary contact switches are used to operate the canopy, one for opening and one for closing. The "OPEN" switch is the push button type. The "CLOSE" switch is a toggle switch with a center off position, a "PARTLY CLOSED" and a "FULL CLOSED" position. The canopy must be partly closed before it can be fully closed. After closing, the canopy is locked by pulling the locking handle (figures 1-8 and 1-9, reference 1), aft. An indicator light (figures 1-6 and 1-7, references 19 and 15) glows when the canopy is closed but unlocked. The exterior wells (figure 1-11) contain switches for normal operation, and a cable to actuate the jettison mechanism in an emergency.

1–74A. ATTITUDE GYRO. A type J-3 attitude gyro is installed in some airplanes and a type A-1 (A-2) or J-8 indicator in others. These instruments provide visual indication of any pitch and roll attitude. They operate on 115V 3 phase AC power supplied by the inverters.

In these instruments the gyro is inclosed in a sphere, a portion of which is visible through the opening of the face of the instrument.

1-74B. The indications of these instruments may be confusing since the presentation of pitch differs.

a. A horizon bar on the A-1 and the J-8 present a conventional pitch indication with the miniature airplane appearing above the horizon bar in a climb and below the horizon bar in a dive. However in a climb (or dive) exceeding 27 degrees of pitch, the horizon bar stops at the bottom (or top) of the instrument case and the sphere then becomes the reference.

Note
The main difference between the A-1 (A-2) and J-8 attitude gyros is that the J-8 has a manual caging control.

b. The J-3 indicator differs from conventional attitude indicators in that climb and dive are not shown in relation to a horizon bar but are read directly on a sphere. The upper hemisphere, which is dark in color, indicates a dive; the lower light hemisphere indicates a climb. Lines similar to latitude markers are painted on the sphere and indicate the amount (degrees) of pitch. In addition a sensitive pitch indicator furnishes readings of climb or dive up to 10 degrees in one degree increments.

Note
The sphere is stabilized maintaining its equator parallel to the earth's surface and the aircraft (and miniature airplane) maneuvers around the stabilized sphere. Therefore when the aircraft is in a nose-high attitude, the miniature airplane will be displaced downward on the light portion of the sphere and in a dive onto the dark portion of the sphere.

CAUTION

In some instances the A-1 (A-2) and J-8 attitude gyros may take as much as 13 minutes to erect itself.

1-75. OPERATIONAL EQUIPMENT.

1-76. Operational equipment including cockpit pressurizing, oxygen, armament, communication, and lighting equipment is described in Section IV.

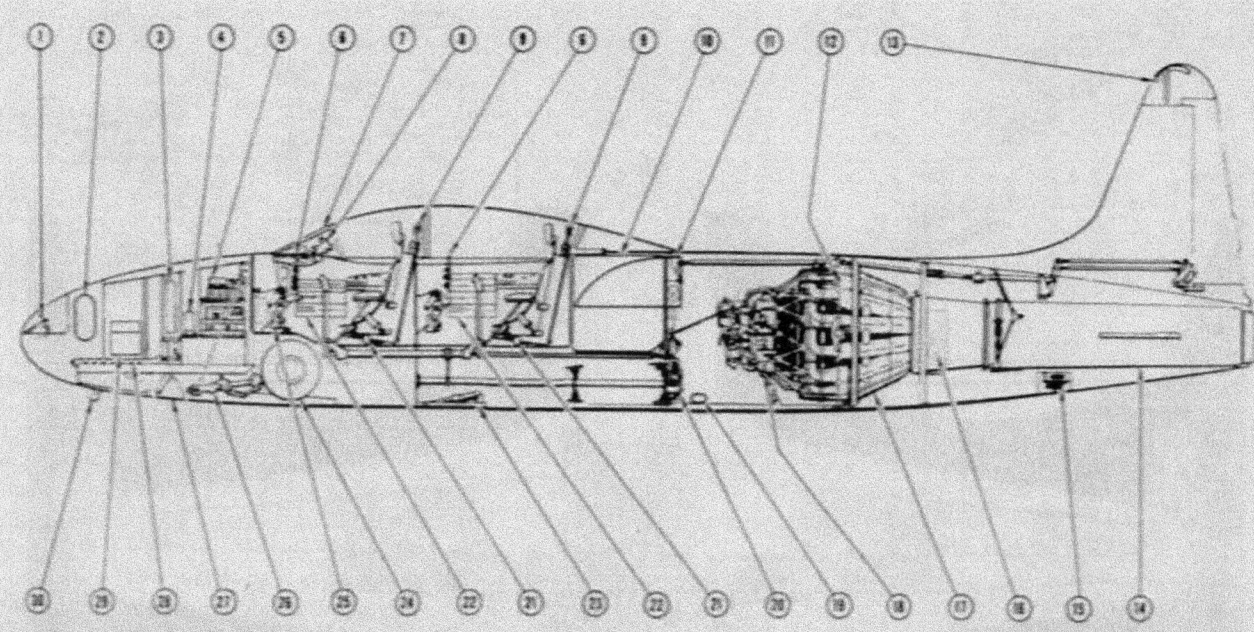

1. AN/ARN-6 Radio Compass Loop
2. Nose Oxygen Cylinder
3. Ammunition Boxes (2)
4. Marker Beacon Receiver
5. AN/ARC-3 & AN/ARN-6 Radios
6. Instrument Panels
7. Bulletproof Windshield
8. Gun Sight
9. Jettisonable Pilots' Seats
10. Fuselage Fuel Tank
11. Water Tank
12. Fuselage Aft Section Attachment
13. AN/ARC-3 Radio Pickaxe Antenna
14. Tailpipe
15. Gyrosyn Compass Flux Valve
16. Elevator Tab Motor
17. Engine
18. Fuel Flowmeter
19. External Power Receptacle
20. Aileron Booster Unit
21. "G" Valves
22. RH Circuit Breaker Panels
23. Dive Recovery Flaps
24. Nose Landing Gear
25. Rudder Pedals
26. Landing-Taxi Light
27. Cartridge Case Ejection Door
28. .50 Caliber Machine Guns (2)
29. Battery
30. Pitot Head

Figure 1-12 — General Arrangement

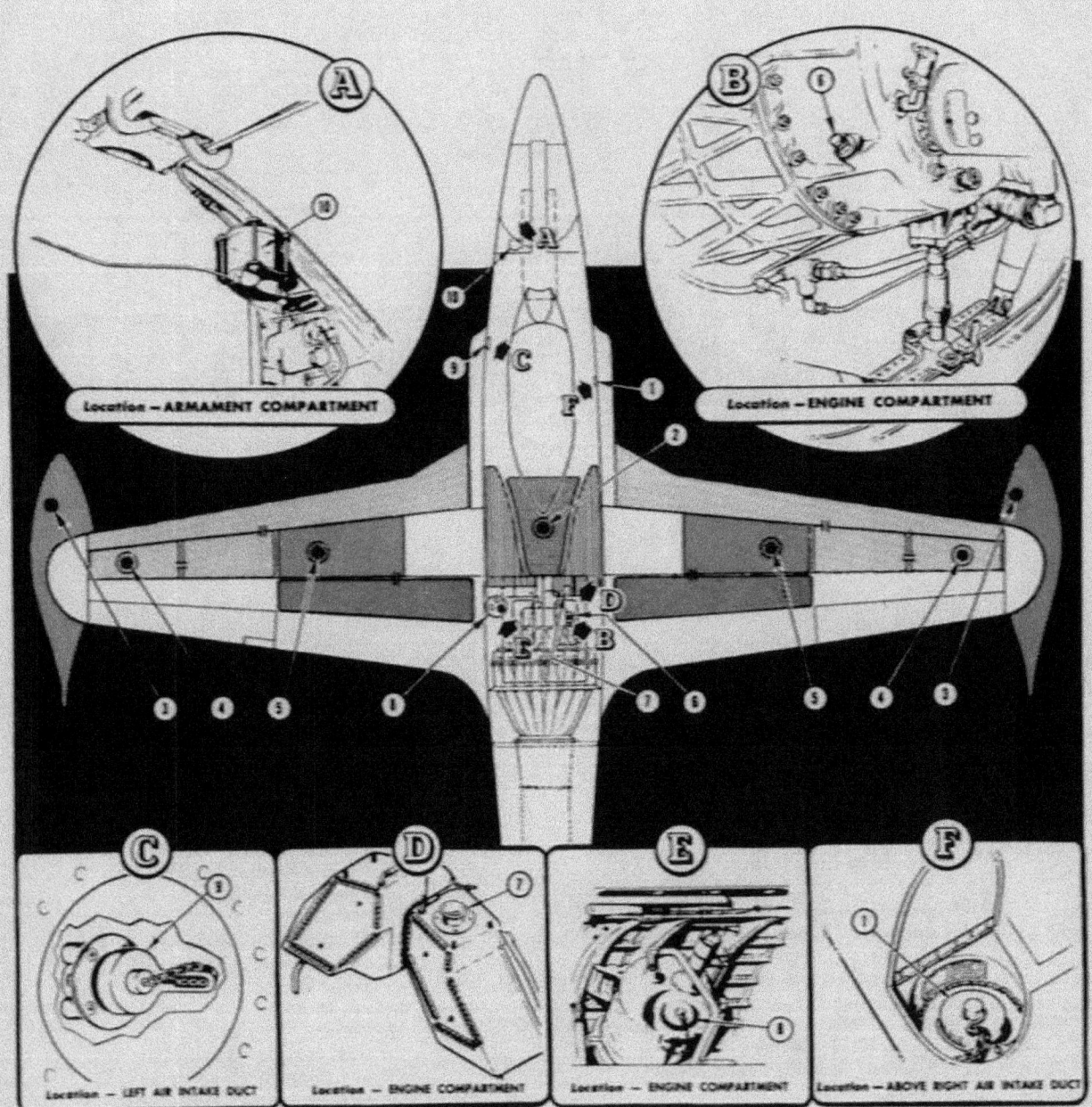

Figure 1-13 — Replenishment Chart

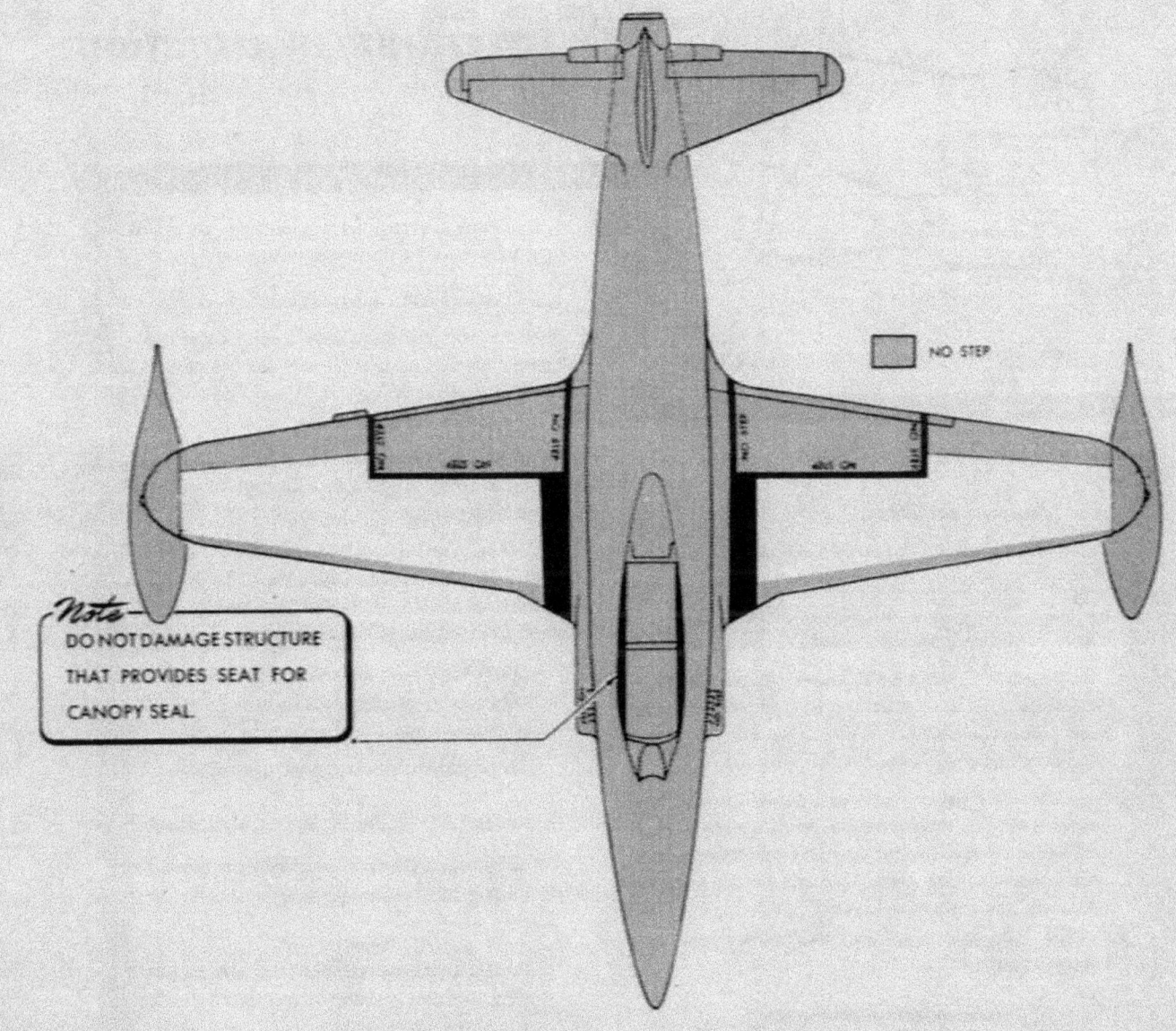

Figure 1-14 — No Step Diagram

SECTION II

NORMAL OPERATING INSTRUCTIONS

2-1. BEFORE ENTERING THE AIRPLANE.

2-2 FLIGHT RESTRICTIONS.

a. Spins are prohibited.

b. Never unlock the canopy in flight.

c. For solo flight, the airplane must be operated from the front cockpit as the rear cockpit does not have complete operating controls.

d. Inverted flying or any maneuver resulting in extended negative acceleration, will result in engine flame-out since there is no means of insuring a continuous flow of fuel in this attitude.

e. Do not attempt to take off with full drop tanks unless the ammunition boxes are full or the equivalent weight (180 lbs.) is carried in the nose armament compartment. If either of the guns are removed, the equivalent weight must be carried (75 lbs. per gun) as ballast in the nose armament compartment.

WARNING

Without proper ballast it is possible to obtain a CG position far aft of the rearward limit.

f. Avoid landing with one drop tank full and one empty. Drop the heavy tank at least.

g. Vertical stalls are prohibited.

h. Aileron rolls with full drop tanks are not recommended, and are prohibited at rates faster than 45 degrees per second (one complete roll of 360 degrees in eight seconds).

i. Release the drop tanks one at a time, in a skid, (left tank first) with the tank to be released on the trailing wing.

j. Do not exceed $+7.33$ or -3 "G." With drop tanks on and full, the following "G" limits will be observed: $+5.33$ in pull-ups; and -2"G" in nose down or inverted maneuvers.

k. Avoid acrobatics and maneuvers involving:
   Large yaw angles at all speeds.
   Violent rolling pullouts at all speeds.
   Uncoordinated turns and steep spirals.

2-3. AIRSPEED LIMITATIONS. (Indicated).

a. Maximum allowable airspeed is .8 mach number or 580 mph indicated whichever is slower.

Note
If aileron compressibility buzz occurs below .8 mach number, limit speed to that at which buzz occurs.

b. Wing flaps extended 100%—200 mph.

c. Wing flaps extended 50%—230 mph.

d. Wing flaps extended 25%—270 mph.

e. Landing gear extended—225 mph.

2-4. ENGINE RESTRICTIONS. Operation above 100% engine rpm and/or 700°C (720°C on J33-A-35 engines) tailpipe temperature is prohibited because of danger of weakening the turbine wheel and causing failure. Operation below 400°C tailpipe temperature with J33-A-23 engines is prohibited (except on approach) because of possible damage to engine.

2-4A. ENGINE SPEED.

a. Overspeeding in excess of 110% rpm for any period of time will require removal of engine for overhaul.

b. Overspeeding from 105 to 110% rpm for any period of time will require a 25 hour inspection to determine engine serviceability.

c. Overspeeding from 101.5 to 105% rpm for more than 15 seconds will require a 25 hour inspection to determine engine serviceability.

d. Overspeeding from 101.5 to 105% rpm for less than 15 seconds will require normal pre-flight inspection to determine engine serviceability.

e. When overspeeding is encountered beyond 101.5% and not in excess of 110% rpm, the cause for overspeeding will be corrected prior to further flight.

These limitations and restrictions are subject to change and latest directives and orders must be consulted.

DO NOT EXCEED 5⅓ "G" IN PULL-UPS WITH DROP TANKS ON FULL, OR 7⅓ "G" WITHOUT DROP TANKS!

DON'T LAND WITH ONE TANK FULL AND THE OTHER EMPTY!

2-5. TAKE-OFF GROSS WEIGHT AND BALANCE. (See Handbook of Weight and Balance AN 01-1B-40).

a. The normal take-off gross weight is approximately 11,800 lbs. The maximum gross weight (with drop tanks full) is approximately 14,250 lbs.

b. The center of gravity position will be near the most forward position at take-off, unless the drop tanks are on and full. That is, approximately, 25% MAC without drop tanks, 29% MAC with full drop tanks, assuming that guns and a full ammunition load are carried.

2-6. EXTERNAL CHECK.

a. Guns—Charged. There are no charging provisions in the cockpit.
b. Armament doors—Locked.
c. Pitot tube cover—Removed.
d. Engine access doors—Fastened.
e. Tip tank filler caps—Screwed tight.
f. External power—Connected.

Note

Connect both cables from an adequate auxiliary power source to the dual receptacle on the airplane to insure that at least 9% rpm will be obtained for starting.

2-7. HOW TO GAIN ENTRANCE.

If a ladder is not available, climb onto the right-hand wing over the leading edge. Operate the switch (figure 1-11) to raise the canopy.

CAUTION

Do not use the gun sight for a hand hold.

2-7A. MINIMUM CREW REQUIREMENT. The minimum crew requirement for this aircraft is one pilot in the front cockpit.

2-8. ON ENTERING THE AIRPLANE.

2-9. CHECK FOR ALL FLIGHTS.

2-9A. FRONT COCKPIT.

a. Weight and balance form F—Check.
b. Forms 1 and 1A—Check.
c. Landing gear lever—"DOWN."
d. Parking brake—Set.
e. Surface control lock—Remove and stow.
f. Oxygen regulator diluter lever—"NORMAL OXYGEN."
g. Oxygen regulator altitude dial—"NORMAL."
h. L.H. Circuit breakers—Reset.
i. Armament switches—Off.
j. Radio—Off.
k. Fuel selector and fluid injection switches—Off.
l. Aileron boost shut-off switch—On.
m. Oxygen pressure—400 to 450 psi.
n. Clock—Set.
o. Fuel counter—Check for proper setting.
p. Battery switch—"OFF."
q. Generator switch—Check "ON."
r. Pitot heater switch—"OFF."
s. Emergency hydraulic pump switch—"OFF."
t. R.H. Circuit breakers—Reset.

2-9B. REAR COCKPIT.

a. Fuel Bypass circuit breaker—Reset (Push in).
b. Jato-Guns switch—"ARM."
c. Bomb Arm and Release switches—"ARM."
d. Fus Tank and Bypass master switch—"NORMAL."
e. Emergency Fuel switch—"OFF."
f. Fluid Injection switch—"OFF."
g. Starting Fuel Sequence switch—"OFF."
h. Landing Light switch—"OFF."
i. Dive Flap switch—UP.
j. Ventilation Control—On.
k. AN/ARN-6 Radio Compass Function switch—"OFF."
l. Cabin Pressurization Inlet Grills and Rear Duct—Shut for solo flights, Open for dual flights.
m. Ignition Normal-Off switch—"NORMAL."
n. Battery and Generator switch—"ON."
o. Emergency Hydraulic Pump switch—"OFF."
p. Aileron Boost Emergency Shut-off switch—"ON."
q. Right Hand Circuit Breakers—Reset (Push in).
r. AN/ARC-3 Radio—Off.

2-10. SPECIAL CHECK FOR NIGHT FLIGHTS.

a. Landing lights and taxi light switch—Test (five seconds maximum for test).

b. Fluorescent lights—Test.

c. Navigation lights—Test.

d. Portable spotlight—Test.

2-11. FUEL SYSTEM MANAGEMENT.

2-12. NORMAL SEQUENCE OF FUEL TANK USE. (See fig. 2-1.)

UNMODIFIED AIRPLANES WITH J33-A-23 ENGINES ONLY.

a. Leading edge tanks for starting, with fuselage tank by-passed.

b. After starting, switch over to JP-1 fuel by turning drop (if carried), wing and fuselage tanks on, and leading edge tanks off.

c. Take-off on fuselage tank only (all other tanks off).

d. After take-off turn drop (if carried) and wing tanks on.

e. Leading edge tanks for purging prior to stopping engine.

J33-A-35 ENGINES ONLY AND MODIFIED AIRPLANES WITH J33-A-23 ENGINES.

a. All tanks on for starting and taxiing out to take-off position.

b. Fuselage tank only for take-off.

c. All tanks on after take-off.

2-13. STARTING THE ENGINE.

2-14. AUTOMATIC START.

WARNING

After any ten hot starts the engine shall be inspected. A hot start is one in which the exhaust temperature exceeds 1000°C (1832°F). The ten hot starts constitute an inspection requirement regardless of the time lapse between the starts and therefore all over temperature operation must be entered in Form 1A.

a. Throttle in "OFF" position.

b. Main fuel valve shut-off switch (later airplanes)—OFF. (Guard down.)

UNMODIFIED AIRPLANES WITH J33-A-23 ENGINES ONLY.

c. Wing and drop tank fuel tank switches in OFF (down) position, fuselage tank switch in "BY-PASS" and leading edge tank switch on (up).

J33-A-35 ENGINES AND MODIFIED AIRPLANES WITH J33-A-23 ENGINES. All fuel tank switches on (up).

d. Ignition "NORMAL-OFF" switch in "NORMAL" position.

e. Starter switch — push to "START" position, hold for three seconds and release. The starter motor should bring the engine up to about 10% speed. Do not attempt to start the engine below 9% speed as it will cause serious damage to the engine. In any instance that the 9% speed cannot be obtained, push the starter switch to the "STOP-START" position and release. Then secure an adequate source of auxiliary power prior to attempting a restart of the engine.

IF THE TAILPIPE TEMPERATURE REACHES 900°C AND STAYS THERE FOR 5 SECONDS SHUT OFF THE ENGINE!

f. At 9 to 10% speed turn the starting fuel switch to "AUTO" position. The engine should start and accelerate automatically to a stabilized speed of approximately 20-25% RPM. This stabilized speed varies with ambient air temperature.

WARNING

If tailpipe temperature reaches 900°C and stays there for five seconds, shut down the engine. If cause is known for the high temperature start, correct it. Repeat start. If 900°C, five second, limit is exceeded shut down the engine. The engine should be checked for malfunction before any further starts.

Note

If ignition does not occur within 10 seconds after the starting fuel switch is turned to the "AUTO" position and released, allow the engine to stop rotating and then check the ignition system before attempting to make another start.

g. At 19-20% (25-35% rpm on J33-A-35) speed place the throttle in the "IDLE" position. This automatically turns off the automatic starting control and allows the engine to operate on the normal engine fuel system.

Note

Do not disconnect the auxiliary power source dual cables until the throttle is moved out of the "OFF" position (with the battery switch in the "OFF" position), or the engine fuel supply will be cut off. If the engine stops when the throttle is opened there is something wrong with the normal engine fuel system. Investigate the difficulty and correct.

Note

It is recommended that the starting fuel switch be turned off by pushing the guard down to avoid the possibility of turning to the "MANUAL" position instead of the "OFF" position. This switch must be in the "OFF" position at all times, except during actual starting and stopping operations. If this switch is left in the "AUTO" position the automatic starting system would be energized whenever the throttle was placed in the "OFF" position, and fuel would be supplied to the engine. If it is left in the "MANUAL" position both the normal and emergency fuel systems would be supplying fuel. Under these conditions there is no governor protection and overspeeding is very likely to occur.

h. UNMODIFIED AIRPLANES WITH J33-A-23 ENGINES ONLY. At not less than 30% speed push fuselage fuel tanks and by-pass switch to the "FUS" (up) position and the leading edge fuel tank switch to the "OFF" (down) position. This changes the engine fuel supply from gasoline AN-F-48 to kerosene AN-F-32 for normal operation of the engine.

CAUTION

In event of restart, pull the tail down to drain the unburned fuel from the tailpipe before attempting a restart.

i. With the throttle in the "IDLE" position check that instruments are in desired ranges.

j. Disconnect the external power source and turn on the battery switch.

2-15. MANUAL START. The engine will normally be started on the automatic system. The manual system will be used only in the event the automatic systems fails to function properly.

WARNING

After any ten hot starts the engine shall be inspected. A hot start is one in which the exhaust temperature exceeds 1000°C (1832°F). The ten hot starts constitute an inspection requirement regardless of the time lapse between the starts and therefore all over temperature operation must be entered in Form 1A.

a. Throttle—"OFF."

b. Main fuel valve switch (later airplanes)—OFF (Guard down).

c. UNMODIFIED AIRPLANES WITH J33-A-23 ENGINES ONLY. Wing and drop tank fuel tank switches in off (down) position, fuselage tank switch in "BY-PASS" and leading edge tank switch on (up).

J33-A-35 ENGINES AND MODIFIED AIRPLANES WITH J33-A-23 ENGINES. All fuel tank switches on (up).

d. Emergency fuel switch—"EMERGENCY."

e. Starter switch—Push to "START" position and release.

Note

Starter will continue to operate after switch returns to neutral position and will automatically cut out when engine speed reaches approximately 15% rpm.

f. Turn starting fuel sequence switch to "MANUAL."

g. At maximum obtainable rpm (not less than 9% rpm), move the throttle rapidly to the wide open position, and as soon as the fuel pressure starts to build up, retard the throttle quickly to "IDLE." As soon as combustion rumble is heard or tailpipe temperature starts to rise, turn the starting fuel sequence switch "OFF."

Note

It is recommended that the starting fuel sequence switch be turned "OFF" by pushing the guard down to avoid the possibility of turning accidentally to the "AUTO" position.

Note

If ignition does not occur within three seconds after the throttle is opened, return the throttle to "OFF" and push the starting switch to the "STOP-START" position and release. Pull the tail down to drain the unburned fuel from the tailpipe before restarting.

h. After the engine starts, adjust the throttle as required to keep the tailpipe temperature below 900° C. Attempt to maintain the temperature between 800° and 900° C until the engine reaches idle rpm.

Note

It may be necessary to pull throttle back beyond the idle position to keep from overheating during the start.

i. Accelerate engine to about 55% rpm.

j. Retard throttle rapidly and at the same time turn emergency fuel switch "OFF" in order to return engine to main fuel system.

CAUTION

Switching from emergency to normal fuel system at low rpm will cause an undesirable surge.

Note

Do not disconnect external power supply until emergency fuel switch is in the "OFF" position or a hot surge may occur in the changeover to the main fuel system.

CAUTION

In event of restart, pull the tail down to drain the unburned fuel from the tailpipe before attempting a restart.

k. UNMODIFIED AIRPLANES WITH J33-A-23 ENGINES ONLY. At not less than 30% rpm push wing, drop and fuselage fuel tanks switches to the on (up) position and the leading edge fuel tank switch to the off (down) position. This changes the engine fuel supply from gasoline AN-F-48 to kerosene AN-F-32 for normal operation of the engine.

l. At idling speed (34% rpm) check that instruments are in desired ranges.

m. Disconnect external power supply and turn the battery switch "ON."

2-16. INSTRUCTIONS IN CASE OF FIRE. Refer to paragraph 3-4.

2-17. WARM-UP AND GROUND TEST.

Note

No warm-up is required. If oil pressure is up and 100% rpm can be obtained the engine is ready for take-off. Gyro instruments will require about 2 minutes from the time the battery switch is turned "ON" to get up to operating speed. However, for an IFR take-off 13 minutes is required before the attitude gyro will give correct indications.

2-18. EMERGENCY FUEL SYSTEM CHECK.

a. Emergency fuel switch—"OFF."

b. Run the engine up to 40-60% rpm.

c. Stop movement of the throttle in this range.

d. Push the emergency fuel check switch and hold. (When the engine changes over to the emergency fuel system, the green and amber emergency fuel indicator lights will come on.

Note

From this point, the pilot can return to the normal system as explained in step e, following, or advance the throttle to determine the maximum power available, if he so desires. However, the tailpipe temperature must be maintained within limits by means of the throttle as the Bendix control is not operating.

e. Release the emergency fuel check switch while rapidly retarding the throttle. This must be done to return the engine to the normal fuel system. (When the engine returns to the normal system the green and amber lights will go out, the red light will stay on.)

2-19. DELETED.

2-20. Check the following:

a. Student lockout indicator — Out (If indicator is glowing, push Lockout Release button.)

b. Aileron and elevator tabs—Check operation and set in neutral position.

c. Dive flaps—Check operation (be sure ground crew is clear of the flaps) and return to the "UP" position.

d. Wing flap—Check operation and set at 31.5°.

e. Surface controls—Check for freedom of operation and proper direction of movement.

f. Altimeters—Set.

g. Inverters (front cockpit only)—Check as follows:

(1) After inverters have operated for half a minute or more, push test switch to "TEST ONLY." The gyro instrument warning light should flick on and off briefly, indicating that the stand-by inverter is operating. If the light continues to glow, the indication is that one of the inverters is not operating.

(2) Release the switch. If the standby inverter was operating, the warning light will flick on briefly, then off, indicating that the number one inverter is operating. If the warning light stays on for three seconds, then goes off, the indication is that the number one inverter is not operating and that the automatic relay has switched back to the standby inverter.

(3) Continued glowing of warning light indicates that both inverters are inoperative.

(4) Continued blinking of the warning light indicates a short circuit in the a-c phase, and the inverter power should be cut off by pulling out the circuit breaker button.

h. Stiff-knee clip—Removed.

2-21. TAXIING INSTRUCTIONS.

2-22. The airplane will start to move when engine speed is increased to about 55% rpm. Speed should be maintained in turns of short radius. It is difficult to start moving with the nose wheel turned sharply or on soft ground. Brakes must be used for steering.

2-23. Taxi time should be cut to the absolute minimum. The fuel consumption while taxiing is about the same in gallons per hour as during maximum range cruising at 35,000 feet.

CUT TAXI TIME TO AN ABSOLUTE MINIMUM!

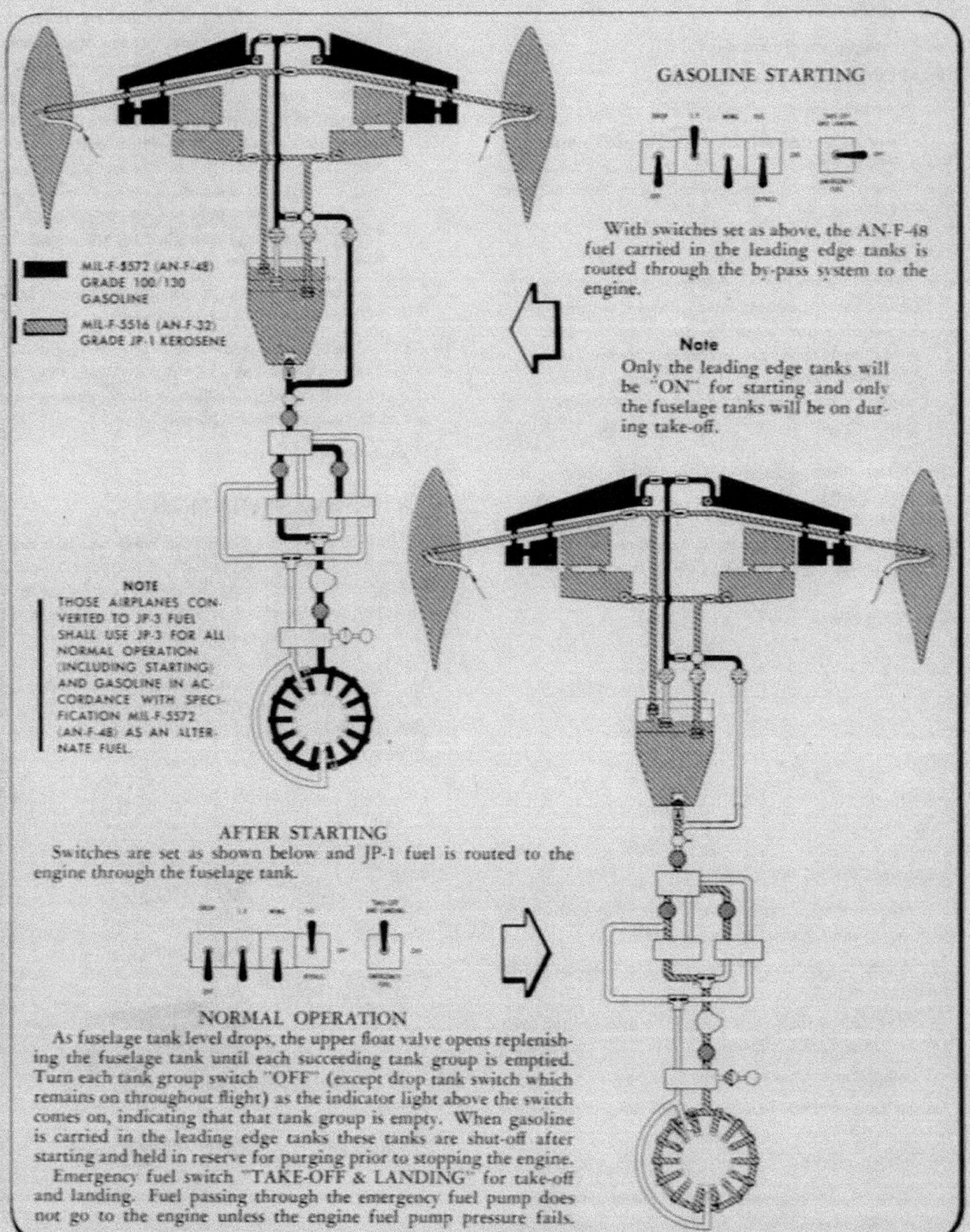

Figure 2-1 (Sheet 1 of 2 Sheets) — Courses of Fuel Flow

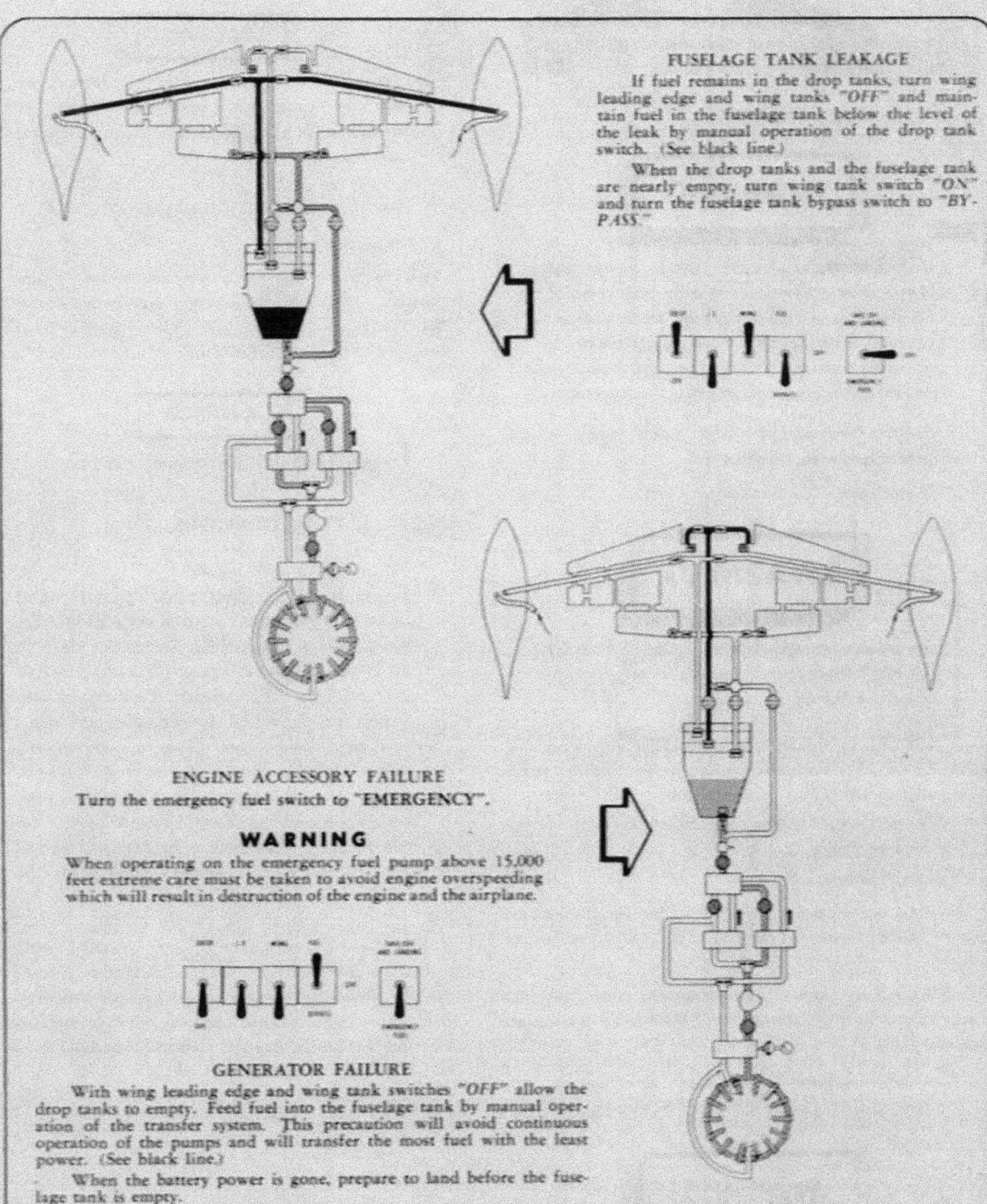

Figure 2-1 (Sheet 2 of 2 Sheets) — Courses of Fuel Flow

Note

A good rule to remember is: Every minute spent on the ground taxiing requires between 3 and 4 gallons of fuel or subtracts about 7 miles from the cruising range of the airplane.

2-24. BEFORE TAKE-OFF.

WARNING

Do not attempt to take off with full drop tanks unless full ammunition or the equivalent weight (180 lbs.) is carried in the nose armament compartment. If either of the guns are removed, the equivalent weight (75 lbs. per gun) must be carried in the nose armament compartment.

a. Shoulder harness and safety belt — tightened and inertia reel lock control unlocked.

b. Wing flaps—Check setting at 31.5°.

WARNING

Do not attempt take-off without this flap setting or length of runway and speed required for take-off will be greatly increased.

c. MODIFIED AIRPLANES WITH J33-A-35 ENGINES ONLY. Turn wing, leading edge and drop tank fuel switches off (down) for take-off.

UNMODIFIED AIRPLANES WITH J33-A-23 ENGINES ONLY. Turn wing, and drop tank fuel switches off (down) for take-off.

d. It is best to use neutral tab if full drop tanks are carried; slight nose up tab if the drop tanks are off or empty.

e. Taxi a few feet straight down the runway so that the nose wheel will be centered when the brakes are released.

f. Canopy—Closed and locked.

g. Emergency fuel switch—"TAKE-OFF and LAND."

WARNING

Check that emergency fuel system green indicator light is on and red and amber lights are out.

h. Gyros—Checked.

CAUTION

In rare instances the attitude gyro may take as much as 13 minutes to erect itself; however in most instances the gyros will erect in three or four minutes.

i. Gyrosyn compass—Synchronize.

j. Hold the brakes.

k. Throttle — Open throttle observing tailpipe temperature. Check the following: Instruments in the desired ranges (see figure A-8, rpm 101.5% maximum, ammeter showing "CHARGE."

CAUTION

Open throttle slowly to prevent flame-out.

2-24A. JATO TECHNIQUE.

Note

Before using JATO you should check the weight and balance to make sure that the static loading limits are not exceeded. Remember that the JATO bottles move the c.g. 2.3% aft. Have the c.g. as far forward as practical because the low thrust line of the JATO power shifts the "effective c.g." further aft. This means that the "nosing up" tendency at the instant of take-off will be more pronounced. You must take prompt corrective action to prevent the tail from striking the runway or to prevent a possible stall, but be careful not to overcontrol because of the light stick forces.

2-24B. Take-off performance will depend somewhat upon the JATO firing point. Minimum ground roll will be obtained when the units are fired shortly after the start of the take-off run, but best performance in clearing a 50-foot obstacle will be obtained by firing the units later in the take-off run.

2–25. TAKE-OFF.

CONSULT THE TAKE-OFF CHARTS IN APPENDIX I !

Note

See paragraph 3–9 for procedure in event of engine failure during take-off.

a. Align the airplane with the take-off runway, apply the wheel brakes, and set the engine speed to 100% rpm.

b. Release the brakes and maintain directional control by minimum use of the brakes until rudder control becomes effective at about 75 mph.

Note

If fluid injection is to be used, set throttle to obtain 98% rpm and turn "ON" fluid injection switch after take-off roll is started. Then adjust throttle to obtain 100% rpm. Check cockpit air for absence of noxious fumes. Complete water-alcohol supply must be used during take-off and initial climb if any part of the flight is to be conducted under conditions where ambient air temperatures will be below —12°C (10°F).

c. As elevator control becomes effective (about 80 mph), lift the nose of the airplane until the nose wheel just clears the runway. In this attitude the total drag is a minimum and the aircraft acceleration will be most rapid.

d. Pull the airplane off the ground at 125 mph with no drop tanks and at 135 mph with drop tanks.

e. Landing gear control—"UP" when definitely airborne.

f. To clear an obstacle in minimum distance, do not allow the airspeed to increase more than 10 mph above take-off airspeed.

g. Wing flaps control "UP" between 160 and 200 mph and completely "UP" by 200 mph.

CAUTION

The airplane has a tendency to sink rapidly when the wing flaps are retracted suddenly; therefore, milk flaps to full up.

h. Climb at about 180 mph to a safe altitude, then accelerate to best climbing speed for the remainder of the climb.

CAUTION

Although it is possible to take off about 10 mph slower than noted above, taking off at too low an airspeed will cause the airplane to settle back on the ground. It must be remembered that sufficient airspeed is important when taking off in this airplane because there is no propeller slip stream to increase the lift of the wing. Also, failure to extend the flaps on take-off will probably cause the airplane to settle back on the ground unless the speeds recommended in paragraph 2–25d are definitely increased.

i. UNMODIFIED AIRPLANES WITH J33-A-23 ENGINES ONLY. Wing and drop tank switches ON (up).

MODIFIED AIRPLANES WITH J33-A-35 ENGINES ONLY. Turn wing, leading edge and drop tank fuel switches on (up).

j. Emergency fuel switch—"OFF."

WARNING

Before turning the Emergency fuel switch "OFF," check to see that the amber emergency fuel indicator light is off. If the amber light is on, the indication is that the main fuel system has failed and the engine is running on the emergency fuel system. If such is the case, turn the emergency fuel switch to the "EMERGENCY" position and circle the field and land.

2-26. CLIMB.

2-27. The speeds for best climb are given in the Climb Chart, appendix I.

2-28. The most economical climb can be obtained at 100% rpm. Do not operate at this power for more than 30 minutes at any one time.

2-29. When supply of water-alcohol has been consumed as indicated by pressure gage dropping to zero, turn the fluid injection switch "OFF."

2-30. DURING FLIGHT.

2-31. The advantage of this airplane lies in its speed. At altitude, its best climbing speed is greater than the top speed of most conventional fighters. The maximum range cruising speed at altitude is also greater than the top speed of some conventional fighters.

DO NOT OVERSPEED THE ENGINE!

2-32. The disadvantage of the airplane lies in its slow acceleration from low speed at altitude. However, once the airplane is in the air there is ordinarily no reason to allow the speed to go below the best climbing speed or the maximum range climbing speed until approaching the field for a landing.

2-33. Possible malfunction of the tip tank fuel system may cause one tip tank to empty and one tip tank to remain full. This condition results in wing heaviness, affecting the wing on which the full tank is installed. Wing heaviness from this cause becomes more apparent as airspeed is reduced and below 114 mph IAS full aileron control and aileron trim will not hold the wings level. Since several fatalities have resulted from attempting to land the subject aircraft in this condition, compliance with the following instructions is mandatory whenever abnormal wing heaviness is encountered and the external wing tip tanks are installed.

a. Jettison the wing tip tanks. Care should be exercised to assure that tip tanks are not jettisoned over congested areas where other lives may be endangered.

b. In the event circumstances beyond the control of the pilot prevent jettisoning the tip tanks and a landing with one full and one empty tip tank becomes necessary, the pilot will make every effort to attain at least 10,000 feet altitude above surrounding terrain and accomplish a simulated landing to determine the lateral control characteristics of the aircraft. Descent from altitude will be made with landing gear and dive brakes (if used) extended and the landing will be accomplished at least 10 mph in excess of the airspeed at which loss of lateral control was noted during the simulated landing.

2-33A. If at any time when carrying wing tip tanks, lateral control and trim become difficult and erratic, reduce airspeed immediately. If the difficulty persists at approximately 200 mph IAS, jettison the wing tip tanks before further investigating the trouble. When lateral control difficulties are encountered, the aileron boost will not be turned off while the tip tanks are still on the aircraft. In the event the wing tip tanks fail to release by normal and emergency means, reduce airspeed to 150 mph and if satisfactory lateral control cannot be maintained, abandon the aircraft. Abandon the aircraft if satisfactory lateral control cannot be maintained after jettisoning the wing tip tanks and shutting off the aileron boost.

2-33B. AILERON. The airplane has a very high rate of roll at any altitude. With tip tanks installed, the lateral stability is such that some attention is required when flying at high altitude, particularly in rough air. Other than frequent reference to lateral attitude, no special technique is required.

2-33C. AILERON BOOST. While there is usually no reason for turning aileron boost off in flight, it can be turned on or off at any time provided the airplane is trimmed laterally and the control stick is centered. If the airplane is not trimmed and the stick is off center, when the boost is shut off, a suddenly increased force will be required to hold the lateral position. Aileron forces without boost increase with air speed. Therefore, in case of failure of the booster unit, it is best to reduce speed as necessary to give lighter aileron forces. Manual aileron control may be supplemented by the use of aileron tab to obtain a fairly high rate of roll at high speeds but it is not considered good technique to depend on the tab for the fine control required such as in formation or other precision flying. It is recommended that the aileron booster be turned off in low speed flight for practice and first hand information on the control forces required.

Note

Due to the tolerance allowed in the governor, the engine may turn up only 98.5% or may turn up 101.5% rpm. Although 100% is the normal full throttle maximum, any value between 98.5% and 101.5% is acceptable so long as the stable tailpipe temperature does not exceed 700° C (Tailpipe temperatures up to 900° C are allowed only while the engine speed is increasing).

WARNING

In event the engine controls allow the above mentioned limits to be exceeded, the pilot should retard the throttle as necessary. Exceeding these limits will adversely affect engine strength and life.

2–34. **STABILITY.** The airplane is directionally and longitudinally stable at all approved center of gravity positions. Laterally the airplane is neutrally stable; therefore, attention is required to hold the wings level when flying in rough air.

Note

With drop tanks installed, the airplane has a reverse rolling tendency when attempting to lift a wing with the rudder. That is, a bank cannot be corrected for by using opposite rudder, but should be corrected for by use of the ailerons.

2–35. **TRIM CHANGES.** Since there is no torque effect from the power plant on this airplane, the rudder forces are zero for all speed and power conditions if the rudder tab is correctly adjusted on the ground.

2–36. The elevator tab should be used with caution, especially at high speeds. Failure of the tab mechanism resulting in excessive trim can be over-controlled by reducing speed.

2–37. The trim change due to lowering the landing gear or flaps or changing power is negligible.

2–38. When the dive flaps are extended at high speed there is a tendency for the nose to come up rapidly. At low speed, this tendency is comparatively slight.

2–39. **CHANGING POWER IN FLIGHT.** With J33-A-23 engines it will be necessary to maintain tailpipe temperature above the lower limit of 400°C. In descending from high altitudes it will probably be necessary to use the dive flaps to hold the IAS down to the recommended descent speed of .6 mach number while using enough power to keep the tailpipe temperature above the lower limit of 400°C.

CAUTION

Open throttle slowly to prevent flame-out.

2–39A. **FLUID INJECTION IN FLIGHT.** Complete water-alcohol injection that is retained for use as thrust augmentation below 10,000 feet during flight or landing, will be utilized as follows:

 a. If used when operating on the emergency fuel system, as in the case of a main fuel pump failure, advance throttle and obtain maximum rpm prior to turning on fluid injection switch. Nearly full dry power can be obtained.

 b. If used when operating on main fuel system as in case of combat training, familiarization, etc. advance throttle to obtain 98 per cent rpm, turn on the fluid injection switch, then adjust the throttle to obtain 100 per cent rpm.

2–40. **STALLS.** (See fig. 2-1A)

2–41. **NORMAL STALLS.** The stall is preceded by noticeable mushing and by buffeting which gives at least 10 mph warning. In a complete stall with power on or off, one wing may drop. If the stick is held back after the stall, the airplane may fall into a steep spiral and may spin.

2–42. Recovery from the stall is made by releasing the back pressure on the stick and lifting the down wing with the ailerons. The rudder is not effective in lifting a dropping wing.

2–43. The stall will occur near the airspeeds indicated in fig. 2-1A at the gross weight noted, however, since it is improbable that a pilot will know his exact gross weight at the time and since the actual stall also depends upon the technique used, it is recommended that stalls be practiced so that they may be anticipated through the feel of the airplane rather than through reference to the airspeed indicator alone.

2–44. Accelerated stalls should be avoided whenever droppable tanks are carried because high loads are imposed on the attachments at high "Gs" and because some airplanes tend to snap roll concurrently with the stall.

2-44A. TURBULENT AIR AND THUNDERSTORM FLYING.

Note

Flight through a thunderstorm should be avoided if it is at all possible. However, since circumstances may force you at some time to enter a zone of severe turbulence, you should be familiar with the techniques recommended for flying the airplane under such conditions.

Power settings and pitch attitude are the keys to proper flight technique in turbulent air. The power setting and pitch attitude required for desired penetration airspeed (figure 2-1B), and established before entering the storm must—if maintained throughout the storm—result in a constant airspeed, regardless of any false readings of the airspeed indicator. Specific instructions for preparing to enter a storm and flying in it are given in the following paragraphs.

2-44B. APPROACHING THE STORM.—It is imperative that you prepare the airplane prior to entering a zone of turbulent air. If the storm cannot be seen, its proximity can be detected by radio crash static. Prepare the airplane as follows:

a. Adjust power controls as necessary to obtain safe penetration speed.

b. Pitot heater—on.

c. Check gyro instruments for proper settings.

d. Safety belt—tightened.

e. Turn off any radio equipment rendered useless by static.

f. At night, turn cockpit lights full bright or use dark glasses to minimize blinding effect of lightning.

CAUTION

Do not lower gear and flaps as they merely decrease the aerodynamic efficiency of the airplane.

2-44C. IN THE STORM.

a. Maintain power setting and pitch attitude (established before entering the storm) throughout the storm. Hold these constant and your airspeed will be constant —regardless of the airspeed indicator.

b. Devote all attention to flying the airplane.

c. Expect turbulence, precipitation, and lightning, and don't allow them to cause undue concern.

d. Maintain attitude. Concentrate principally on holding a level attitude by reference to the artificial horizon.

e. Don't chase the airspeed indicator, since doing so will result in extreme airplane attitudes. If a sudden gust should be encountered while the airplane is in a nose high attitude, a stall might easily result. A heavy rain, by partial blocking of the pitot tube pressure head, may decrease the indicated airspeed reading by as much as 70 mph.

f. Use as little elevator control as possible to maintain your attitude in order to minimize the stresses imposed on the airplane.

g. The altimeter is unreliable in thunderstorm flying because of differential barometric pressures within the turbulent area. A gain or loss of several thousand feet may be expected. Make allowance for this error in determining minimum safe altitude.

Note

Normally, the least turbulent area in a thunderstorm will be at an altitude of 6000 feet above the terrain. Altitudes between 10,000 feet and 20,000 feet are usually the most turbulent.

2-45 SPINS

2-46. Intentional spins are prohibited due to the fact that the airplane does not always spin in the same manner and considerable altitude may be lost before final recovery, particularly in the gear and flap down configuration. Should the airplane inadvertently enter a spin, immediate recovery should be made by neutralizing the controls or by using the rudder and aileron against the spin. If any difficulty is experienced, the following procedure should be used: When the airplane reaches a nose down attitude during the gyrations of the spin, the controls should be utilized to keep the nose down and

STALLING SPEEDS — IAS - MPH

Gross Weight	Approx. Fuel Remaining	GEAR & FLAPS UP			GEAR & FLAPS DOWN		
		Level Flight	30° Bank	60° Bank	Level Flight	30° Bank	60° Bank
14,000 lbs.	680 gals.	130	140	185	115	125	165
12,000 lbs.	350 gals.	120	130	170	105	115	150
10,000 lbs.	50 gals.	110	120	155	95	105	135

Figure 2-1A — Stall Speed Table

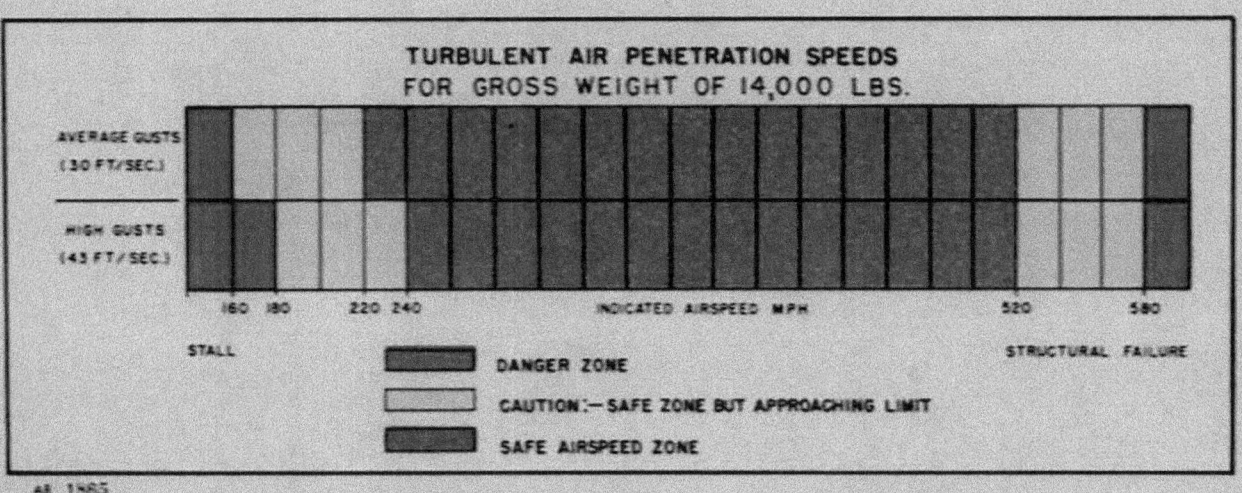

Figure 2-1B — Turbulent Air Penetration Speed

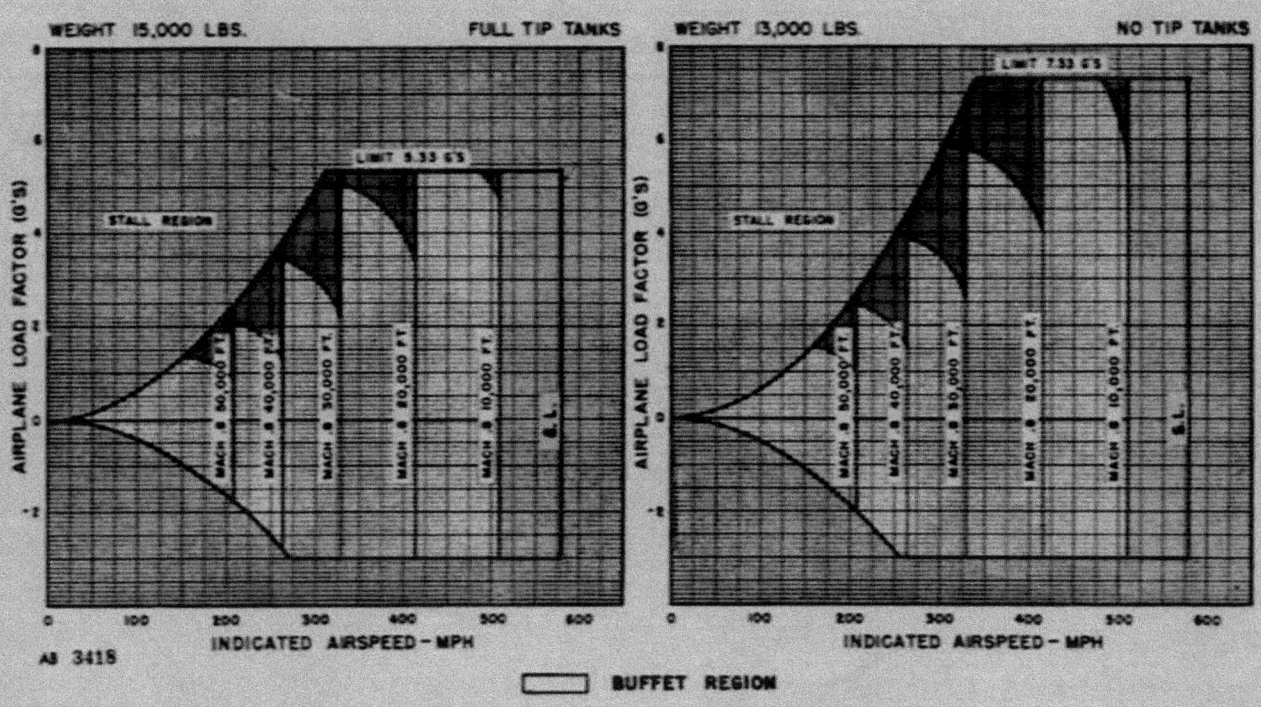

Figure 2-1C — Load Factor versus Airspeed Diagram

increase the airspeed. As the airspeed increases, recovery should be made, taking care not to restall the airplane with the possibility of reentering the spin.

PARAGRAPHS 2–47 THRU 2–54 DELETED.

Figure 2-2 — Spin Pattern

2–55. INVERTED SPINS.

2–56. CHARACTERISTICS. Inverted spin characteristics, so far as known, are similar to those of the normal spin except that flame-out will occur, due to the inverted attitude, and the spin has a tendency to go from the inverted into a normal spin.

2–57. RECOVERY. The recommended recovery procedure is to neutralize the controls or use rudder and aileron against the spin if desired and when rotation has stopped, half roll to the normal attitude and pull out, using caution to avoid an accelerated stall.

2–58. PERMISSIBLE ACROBATICS.

2–59. Acrobatics, except those requiring extended negative acceleration, are permissible. Under negative acceleration conditions, fuel will not be fed to the engine and flame-out will occur if the inverted condition is maintained for more than a few seconds.

2–60. Ten quarts of oil are required in the engine reservoir to provide sufficient lubrication during acrobatics.

2–61. The pilot is cautioned to use extreme care in maneuvers which require a downward recovery as the loss of altitude in downward recovery is very rapid. In general, acrobatics should not be attempted below 10,000 feet until the pilot becomes familiar with the speed at which the airplane can gain and lose altitude.

2–62. DIVING.

2–63. The airplane is controllable up to a mach number of .8. These limitations must be observed. At the critical mach number, lateral control is very difficult and uncertain even though longitudinal control is still good. Aileron buzz may occur slightly before, or at, the speed at which lateral instability is noticed.

2–64. If the use of the trim tab is neglected, considerable push on the control stick will be required to hold the airplane in the dive. This stick force increases up to a mach number of about .75 and will remain approximately constant between mach numbers of .75 and .8.

2–65. When the dive flaps are extended at high speed, there will be a definite nose-up tendency. The acceleration, however, will not be excessive even with "hands off." This nose-up tendency may be counteracted by applying nose-down trim tab at the same time extension of the dive flaps is started.

2–66. Caution must be observed when retracting the dive flaps at high speed, as a sudden nose-down tendency is created which must be resisted if flying close to the ground.

CAUTION

Aileron compressibility "buzz" is a low amplitude vibration of the ailerons which can best be detected by watching for a fuzzy outline at the trailing edge of the aileron. This buzz will occur at about .8 mach number in "one G" flight; slower under accelerated flight condition. Operation within the buzz region should be avoided whenever possible.

2–67. The dive flaps may be extended at any time and at any speed. It is recommended that dives be conducted with the dive flaps up so that they will be in reserve to aid in reducing speed when, and if, trouble is encountered.

2–68. NIGHT FLYING.

2–69. For taxiing at night, place the landing light and taxi light switch in the "TAXI" position and for take-off and landing in the "LANDING LIGHT" position. Be sure the lights are turned off as soon as the gear is retracted to prevent them from burning out.

2-70. APPROACH AND LANDING.

2-71. The landing technique is similar to that of conventional tricycle landing geared airplanes, and the landing attitude is about the same; that is, main wheels first, tail slightly down.

Note

Care should be exercised to avoid a tail-low attitude which will cause the tail to drag on the runway.

2-72. When landing in headwinds, the approach speed should be increased 5 mph for each 10 mph of reported headwind, due to possible gust conditions. This increase in speed should be obtained by application of power. Such a procedure will result in a flatter approach angle, thereby reducing gust effects.

CAUTION

If wing heaviness due to uneven fuel transfer from the drop tanks should be encountered, it is strongly recommended that the heavy tank be dropped before landing.

2-73. With the landing gear down and wing flaps 100% extended, start the final turn at about 150 mph indicated airspeed. When the landing is assured, start flaring off. Come over the end of the runway at 120 mph if the airplane has about 100 gallons of fuel remaining. If the landing is being made with an appreciable load of fuel or ammunition, the above airspeed should be increased about 3 mph for each additional 100 gallons.

2-74. Keep the engine at 50% to 60% rpm during the approach so that the power may be applied more quickly if it should become necessary to go around. Approximately 20 gallons of fuel will be required for a tight pattern on the go-around for landing.

2-75. If, for some reason, the flaps cannot be lowered, land aproximately 20 mph faster and allow for more flare-off and a much flatter gliding angle.

2-76. Dive flaps may be used as desired during the approach and landing. Their use will increase the glide angle and slightly reduce the length of roll after landing.

2-77. NORMAL LANDING PROCEDURE.

   a. Safety belt and shoulder harness—tightened and inertia reel lock control unlocked.

   b. Landing gear—"DOWN" (not over 225 mph).

   c. Emergency fuel switch—"TAKE-OFF and LAND."

Note

In those aircraft where the emergency fuel switch is marked "TAKE-OFF"–"OFF"–"EMERGENCY" place switch in the "OFF" position for landing.

Note

Side slips, fish tailing, and "S" turns may be used as desired. These maneuvers should be practiced in normal landings so that they may be used more effectively in case of an emergency "dead-stick" landing.

   d. Wing flaps—"DOWN" (not over 200 mph).

   e. Dive flaps—"DOWN" (if desired).

   f. Engine speed—50% to 60% rpm.

   g. Wing and dive flaps—"UP" before taxiing.

   h. Emergency fuel switch—"OFF."

CAUTION

Excessive use of the brakes must be avoided. As a rule, braked landings should not be made oftener than once every 15 minutes. Heat generated by too much braking will cause tire failure.

2-78. CROSS WIND LANDING. Same as a normal landing. If the drift appears excessive, the upwind wing may be lowered until just before contact.

2-79. MINIMUM RUN LANDING. Follow normal landing procedure and try to land as close as possible to the edge of the field. Go over the fence low and use dive flaps in addition to the wing flaps. After contact, use the brakes as much as possible without sliding the tires.

2-80. EMERGENCY LANDING. See paragraph 3-33.

2-81. TAKE-OFF IF LANDING IS NOT COMPLETED. The ability of this airplane to take off in the event the landing is not completed is definitely inferior to that of conventional single engine fighters. If the landing cannot be completed, the decision to go around should be made as early as possible. The refused landing should be made as follows:

   a. Open the throttle to 100% rpm.

CAUTION

Open throttle slowly to prevent flame-out.

   b. Retract the landing gear immediately as soon as safe flying speed is reached.

   c. Milk the flaps up to 50% until the airspeed indicates over 140 mph, then retract them all the way.

TYPICAL PATTERN FOR 100 GALLONS OF FUEL REMAINING
(Add 3 mph to approach speeds for every 100 gallons above this figure)

Figure 2-3 — Approach Diagram

d. Accelerate to approximately 165 mph before starting to climb.

2-82. STOPPING THE ENGINE.

MAINTAIN ADEQUATE SPEED ON APPROACH. JET AIRPLANES DON'T ACCELERATE AS RAPIDLY AS THOSE WITH PROPELLERS!

2-83. STOP THE ENGINE as follows:

a. Set parking brakes.

b. (With J33-A-23 engines only.) Upon parking the airplane move the throttle to the "IDLE" position, push the leading edge fuel tank switch to the "L.E." (up) position; the fuselage fuel tank and by-pass switch to the "BY-PASS" position; and the remaining fuel tank switches to the "OFF" position. Push the starting fuel switch to the "AUTO" position. Allow the engine to idle for two minutes on gasoline and pull the throttle to the "OFF" position. The engine will continue to run on the starting control. After ½ minute operation push the starting fuel switch to the "OFF" position. This procedure will purge the engine fuel system of kerosene AN-F-32 providing gasoline for the next start.

c. Throttle—"OFF."

d. Turn all switches "OFF" except generator switch.

2-84. BEFORE LEAVING THE AIRPLANE.

2-85. Accomplish the following:

a. Lock the surface controls.

b. Release the parking brakes after chocks are in place.

c. Close the canopy.

SECTION III
EMERGENCY OPERATING INSTRUCTIONS

3-1. EMERGENCY EXIT.

3-2. DELETED.

a. If the airplane is still controllable reduce airspeed to less than 200 mph.

b. Disconnect oxygen and radio equipment and shoulder harness.

CAUTION

If bail-out is made at high altitude, remain connected to the airplane oxygen system while all other preparations for leaving the airplane are being made. Just before leaving the airplane, disconnect oxygen mask from mask-to-regulator tubing and place the type H-2 emergency oxygen cylinder into operation, by pulling the rip cord cable of the oxygen cylinder (the caution tag and pin assembly having been removed prior to take-off).

c. Leave canopy locking handle in the locked position.

d. Jettison the canopy by pulling down on the jettison lever in either cockpit.

WARNING

Bend forward and lower the head when jettisoning the canopy to avoid injury from the released canopy.

e. Release the safety belt.

f. Crawl out on either side or roll the airplane on its back and push clear of the airplane.

3-3. DELETED.

3-4. FIRE.

3-5. There is no fire extinguishing system on this airplane. If the overheat warning light comes on, reduce power to see if the light will go out, especially if the engine was operating at high power.

3-6. If the light goes out when power is reduced, continue flight with caution and land as soon as possible.

3-7. If the light does not go out, or if the fire warning light comes on, shut down the engine completely (fuel switches and throttle off) and on later airplanes turn the main fuel shut-off valve switch "ON."

3-8. Make reasonably sure that fire is actually present before abandoning the airplane, as described in paragraph 3-1.

3-8A. COCKPIT SMOKE REMOVAL.

Note

During certain atmospheric conditions the air-cooler will create a vapor condition which resembles smoke. This can usually be eliminated by temporarily moving the cockpit temperature control to the full hot position. If the condition persists, shut off the pressurization air by closing the pressurization grills and overhead ducts.

a. Turn off the battery and generator switch until it is determined that the smoke is not caused by electrical wiring. See paragraph 3-24 for effects of electrical failure.

b. Open the cockpit ventilation controls (fresh air blast tube on the left).

ENGINE FAILURE

c. Close the front and rear cockpit pressurization grills and overhead inlet ducts. (The rear ducts are closed on the preflight check if the airplane is to be flown solo.)

d. Extend the dive flaps and open the dump valve.

CAUTION

When flying above 37,000 ft. with full pressurization, the dump valve should be opened intermittently until cockpit pressure is reduced to zero, to avoid the effects of explosive decompression. After pressure is released, check to see that the dump valve lever stays in the down (dump) position.

3-9. ENGINE FAILURE DURING TAKE-OFF.

3-10. TOTAL POWER FAILURE BEFORE LEAVING THE GROUND.

a. Throttle—"OFF" immediately.

b. Use the brakes as required.

c. Drop tanks or bombs—Release if it is necessary to retract the gear.

d. Landing gear—Retract if there is insufficient runway.

Note

If the airplane is still on the ground it will be necessary to release the landing gear lever lock by depressing the down lock release lever.

e. Battery and generator switches—"OFF."

3-11. TOTAL POWER FAILURE AFTER LEAVING GROUND. If total power failure occurs soon after leaving the ground, accomplish as much of the following as conditions permit.

a. Throttle—"OFF."

b. Drop tanks or bombs—Release.

c. Land straight ahead.

d. Landing gear—"UP" if it is not possible to land on the runway.

e. Wing flaps—Leave extended.

f. Battery and generator switches—"OFF" before ground contact.

3-12. PARTIAL POWER LOSS. If the engine rpm should drop off, at any time during a take-off, the first thing to do is to make a decision whether to go around or to stop the airplane on the ground.

3-13. POWER LOSS BEFORE LEAVING GROUND. If partial power failure occurs on the ground, proceed as in paragraph 3-10.

3-14. POWER LOSS AFTER LEAVING GROUND. If the airplane is already airborne and partial power failure occurs:

a. Throttle—Wide open.

b. Fluid injection switch—"ON" (after engine has accelerated above 90% rpm) if water is available.

c. Drop tanks or bombs—Release.

d. Landing gear—"UP."

e. Push the nose of the airplane down as much as necessary to obtain a constant increase in airspeed.

f. Start to milk the flaps at 135 mph.

g. When sufficient speed and altitude have been obtained, circle the field and land.

3-15. ENGINE FAILURE DURING FLIGHT.

CAUTION

Aileron boost failure may be expected shortly after engine failure as the engine is the source of hydraulic pressure. However, a by-pass is incorporated in the booster unit to permit manual operation of the ailerons with slightly increased stick forces. See paragraph 2-33C regarding aileron boost failure.

3-16. As soon as flame-out occurs, place throttle in "OFF" position. Immediately determine if fuselage tank contains fuel. If not, start transfer of fuel if it is available.

3-17. Glide down to 25,000 feet (at higher altitudes, poor flame propagation makes air starts very uncertain) before attempting an air start. If circumstances permit, keep engine windmilling speed up to 10% or more. Engine speed should stay above 10% if the air speed is maintained within about 30 knots of the red needle. If circumstances make a fast descent undesirable, the starter may be used as explained in the starting procedure.

3-18. Turn off unnecessary electrical equipment to conserve battery power for starting.

3-19. Air starts should be accomplished by use of the automatic fuel starting system. The manual system should be used only in case of failure of the automatic system.

Note

In case the recommended procedure has been forgotten, the normal ground start procedure will work if the flame-out was due to anything other than a failure of the normal engine pump or engine fuel control.

3-20. AIR START — AUTOMATIC.

a. Pull up for 5 to 10 seconds at 1 G to permit drainage of fuel from tailpipe and combustion chambers. Then hold air speed at about 200 to 225 mph for the start.

b. If engine speed is below 10% rpm, push starter switch to "START" and release. If rpm is 10% or more, omit this step.

c. Air start ignition switch—"START" (and release). Ignition will continue for approximately 45 seconds.

d. Starting fuel sequence switch—"AUTOMATIC" at not less than 10% rpm.

e. Emergency fuel switch—"TAKE-OFF and LAND."

f. After burners light and engine has stabilized on the starting control, open throttle with smooth positive force to idle detent.

g. Turn starting fuel sequence switch "OFF" immediately after setting throttle in idle. If the rpm starts to drop off, open throttle sufficiently to maintain a speed equal to the stable speed on the starting control.

WARNING

1. If the tailpipe temperature reaches 1000°C and stays there for more than three seconds, turn starting fuel switch "OFF" immediately and then move the throttle to the "OFF" position.

2. If the amber emergency fuel indicator light remains on after the throttle is opened, the engine is running on the emergency fuel system. Therefore, leave the emergency fuel switch in the "TAKE-OFF and LAND" position until the airplane is landed. Use extreme care in throttle manipulation to prevent engine overspeeding, engine blowouts, or excessively low engine idle speeds as there is no governor in the emergency fuel system. If the amber light is out, the emergency fuel switch may be returned to the "OFF" position after the throttle has been opened.

h. Accelerate to desired rpm. Note Warning (2).

3-20A. AIR START — MANUAL.

a. Pull up for 5 to 10 seconds at 1 G to permit drainage of fuel from tailpipe and combustion chambers. Then hold air speed at about 200 to 225 mph for the start.

b. If engine speed is below 10% rpm, push starter switch to "START" and release. If rpm is 10% or more, omit this step.

c. Air start ignition switch—"START" (and release). Ignition will continue for approximately 45 seconds.

d. Starting fuel sequence switch—"MANUAL" at not lss than 10% rpm.

e. Emergency fuel switch—"EMERGENCY."

f. Rapidly open throttle to approximately the three-quarters open position. As soon as the fuel manifold pressure begins to rise from zero, rapidly retard the throttle to approximately one inch below the idle detent and place hand on the starting fuel system switch.

g. At indication of flame (sound or temperature), turn the starting fuel system switch to "OFF" and allow engine speed to stabilize.

WARNING

1. If the tailpipe temperature reaches 1000°C and stays there for more than three seconds, move the throttle into "OFF" immediately.

2. Since the engine is operating on the emergency system, use extreme care in throttle manipulation to prevent engine overspeeding, engine blowouts or excessively low engine idle speeds.

h. After engine speed stabilizes (at approximately 25%), slowly advance the throttle lever to obtain desired rpm.

i. If engine flame-out was not due to failure of the main engine pump or main fuel control, engine operation may be returned to the main system by advancing the rpm to about 90 to 100% and then retarding the throttle (quite rapidly) at the same time the emergency fuel switch is moved to the "OFF" position.

3–21. FUEL SYSTEM EMERGENCY OPERATION.

3–22. ENGINE FAILURE. If the engine fails for no apparent reason, it is probable that one of the engine fuel system parts has failed. The engine will run on the emergency fuel system after a normal air start.

3–23. LEAKING FUEL TANKS. It is not probable that leaking tanks will be detected during flight. If a serious leak is suspected, use the fuel from the leaking tank as rapidly as possible (by turning all other tanks "OFF"). If the leak is in the fuselage tank, first turn "OFF" the wing tank and leading edge tank switches and consume the fuel in the drop tanks and the fuselage tank. Then, turn the fuselage switch to "BY-PASS" and the wing tank and leading edge tank switches to "ON."

3–23A. Tip Tanks Fuel System Malfunction.

3–23B. Due to malfunction of the wing tip tank fuel system, it is possible for one tip tank to empty and one tank to remain full. If this occurs, it will result in wing heaviness which will become more apparent as airspeed is reduced and below 114 mph IAS full aileron control and trim will not hold the wings level. Therefore, whenever wing heaviness is encountered and wing tip tanks are installed observe the following instructions:

  a. Jettison the tip tanks.

  b. In event landing with one full and one empty tip tank becomes absolutely necessary, attain at least 10,000 feet altitude above the surrounding terrain and accomplish a simulated landing to determine the lateral control characteristics of the aircraft. Descent from altitude will be accomplished with landing gear extended and the landing will be made at least 10 mph in excess of the airspeed at which loss of lateral control was noted during the simulated landing.

3–23C. If at any time when carrying tip tanks, lateral control and trim becomes difficult and erratic, reduce airspeed immediately. If the difficulty persists at approximately 200 mph IAS, jettison the wing tip tanks before further investigating the trouble. When lateral control difficulties are encountered, the aileron boost will not be turned off while the tip tanks are still on the aircraft. In the event the wing tip tanks fail to release, reduce airspeed to 150 mph and if satisfactory lateral control cannot be maintained, abandon the aircraft. Also abandon the aircraft if satisfactory lateral control cannot be maintained after jettisoning the wing tip tanks and shutting off the aileron boost.

3–24. ELECTRICAL FAILURE.

3–25. COMPLETE FAILURE. If the electrical system should fail completely, fuel will be available only from the drop tanks and the fuselage tank.

CAUTION

In event of an electrical failure the instruments will fall with the pointers remaining in the operating range.

3–26. GENERATOR FAILURE. If only the generator fails and battery power is still available, turn off all unnecessary electrical equipment, and turn wing tanks "OFF" allowing fuel to transfer from the drop tanks. When the drop tanks are empty, turn "ON" the wing tanks until the fuselage tank quantity gage reaches 85 gallons, turn "ON" the wing tanks intermittently to maintain this level until the leading edge tanks are emptied. Repeat this manual transfer from the leading edge tanks.

3–27. EMERGENCY SALVO SWITCH.

3–28. A bomb salvo switch (figures 1-6 and 1-7, references 40 and 38) is located on the left hand side of the sub-instrument panel. Push the salvo switch to release bombs or drop tanks in an emergency.

3–29. WING FLAP EMERGENCY OPERATION.

3–30. Either of the two wing flap motors will extend the flaps. If both motors should fail, or in case of electrical failure, the airplane must be landed with flaps up.

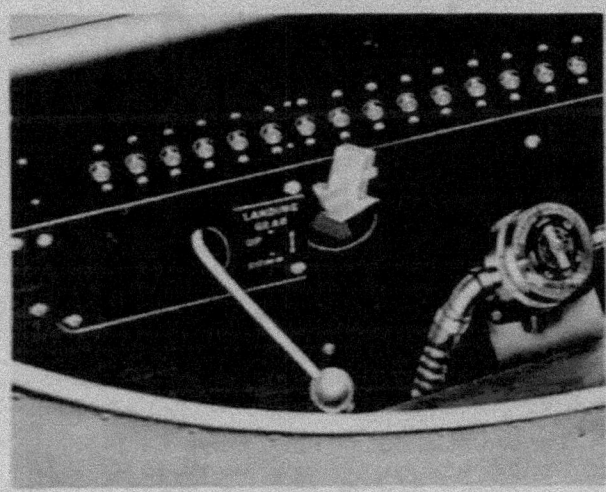
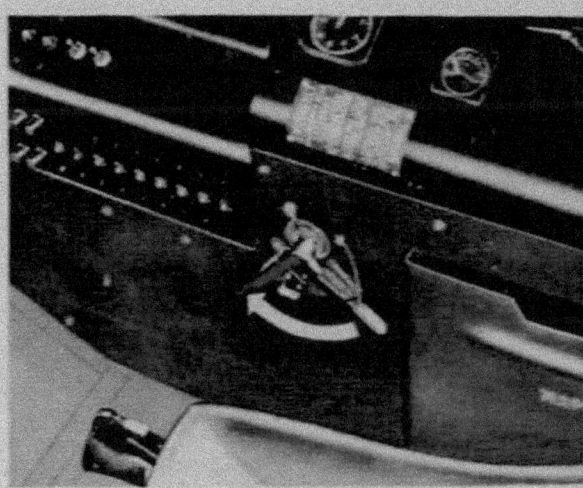

(LH SIDE) NORMAL — (RH SIDE) EMERGENCY

Figure 3-1 — Landing Gear Controls

3–31. LANDING GEAR EMERGENCY OPERATION.

3–32. Proceed as follows, using the emergency hydraulic system:

a. Put the landing gear control in the "DOWN" position.

b. Break the safety wire and turn the emergency hydraulic selector valve lever (figure 3-1) to "EMERGENCY."

c. Turn on the "EMERGENCY HYDRAULIC PUMP SWITCH" until the landing gear is down and locked, as indicated by the position lights.

Note

Do not operate the emergency pump switch until the emergency selector valve lever is placed in the "EMERGENCY" position, as the fluid will only be pumped back to the emergency tank. Recheck the position of the selector lever if results are not obtained.

WARNING

When the gear is extended by use of the emergency hydraulic system, it cannot be retracted again. If the normal hydraulic system failure was due to a break in a line, probably all of the normal system fluid will be pumped overboard during the process, thus aileron boost and dive flaps will not be available.

3–33. LANDING WITH WHEELS RETRACTED.

3–34. For a belly landing proceed as follows:

a. Release the drop tanks. (Push "BOMB SALVO" button.)

b. Jettison the canopy.

c. Make sure that the parachute is unbuckled and that the shoulder harness and safety belt are locked.

CAUTION

The pilot is prevented from bending forward when the shoulder harness lock control is in the locked position; therefore, all switches not readily accessible should be cut before moving the control to the locked position.

Extend full wing flaps (full flaps will prevent wing tip from digging into the ground with resultant ground loops).

d. Move dive flap switch to "UP."

e. Before contact with the ground, move the throttle to "OFF" and turn the generator and battery switches "OFF."

f. Make a normal approach at 10 to 15 mph above the stalling speed and let the airplane touch the ground slightly before the stall is reached.

3-35. LANDING IN WATER. (Ditching.)

BAIL OUT RATHER THAN ATTEMPT A WATER LANDING!

3-36. GENERAL. When anticipating an emergency due to lack of fuel, do not descend near the water to check conditions. The fuel remaining in the airplane will give at least 2½ times more range at 35,000 feet than it will at sea level. Stay at altitude until the fuel is gone, then glide down to a reasonable altitude and bail out.

WARNING

In all cases, it is recommended that the pilot bail out rather than attempt a water landing, if sufficient altitude is available.

3-37. DITCHING. If there is insufficient altitude for a safe bail-out, ditch as follows:

a. Release full or nearly full drop tanks.

Note

Empty or nearly empty tanks will hold ducts out of water until initial speed is lost.

b. Jettison the cockpit canopy.

c. Make sure the landing gear is up.

WARNING

Do not attempt a water landing with the landing gear extended.

d. Unbuckle the parachute harness.

e. Make sure the shoulder harness is locked and the safety belt is fastened.

CAUTION

The pilot is prevented from bending forward when the shoulder harness lock control is in the locked position; therefore, all switches not readily accessible should be cut before moving the control to the locked position.

f. Throttle—CLOSED.

g. Dive flaps "DOWN," wing flaps ½ to ⅔ down.

Note

The wing and dive flaps will not cause the airplane to dive. Open dive flaps will keep jet intakes up.

h. Select leading parallel to wave crest if possible. Aim to touch down on side of crest or on falling side of wave, never on rising side. Land as gently as possible.

i. After airplane comes to rest, get out of the cockpit immediately. Don't forget your life raft.

3-38. HYDRAULIC SYSTEM EMERGENCY OPERATION.

3-39. AILERON BOOST FAILURE. To prevent further failure or sudden recovery of aileron boost while at low altitude, turn the aileron boost switch (figures 1-4 and 1-5, references 20 and 15) to "OFF" at altitude if necessary and at low altitude at all times.

3-40. LANDING GEAR. If the emergency hydraulic system will not extend the gear, replace the emergency hydraulic selector lever in its normal position and place the landing gear control in the "DOWN" position. After these settings are made, hold the hydrofuse reset knob down until either the gear is down and locked as indicated by the green light, or until all the hydraulic fluid is pumped overboard as indicated by failure of the aileron booster. If the landing gear still fails to extend, try the emergency system again before making a belly landing or bailing out.

SECTION IV
OPERATIONAL EQUIPMENT

4-8. PRESSURE SHUT-OFF VALVES. Toe operated shut-off valves (figures 1-6 and 1-7, references 28 and 34) on the grills, one adjacent to each rudder pedal, and a control lever on the inlet duct located adjacent to each pilot's seat turn the cabin pressurizing air on and off.

4-1. VENTILATING.

4-2. DESCRIPTION. Outside air is supplied to the cockpit through a scoop in the left engine intake duct. The air enters the cockpit through tubes which direct it to each pilot's face. A check valve in the tube prevents loss of cabin pressure regardless of the position of the ventilation control lever.

4-3. VENTILATION CONTROLS. A lever near the end of the tube controls the amount of flow and a swivel fitting with a locking ring on the end of the tube directs the flow.

4-4. HEATING AND PRESSURIZING.

4-5. DESCRIPTION.

4-6. GENERAL. Hot air for heating and pressurizing is obtained from the engine compressor. Any portion of this hot air can be cooled by directing it through a turbo refrigerator. The portion of air directed through the turbo refrigerator is later mixed with that portion of the hot air which by-passes the turbo refrigerator, before the air enters the cockpit. (See figure 4-1). Cabin pressure is indicated by the altimeter on the right shelf.

4-7. COCKPIT HEAT CONTROL. The heat control (figure 1-4, reference 17) is located on the left control shelf. Any position between "HOT" and "COLD" may be selected to control the cabin temperature. This control regulates the air temperature regardless of whether or not the cabin is being pressurized, by controlling the amount of air being directed through the turbo refrigerator. The resultant air enters the cockpit through the pressure shut-off valves.

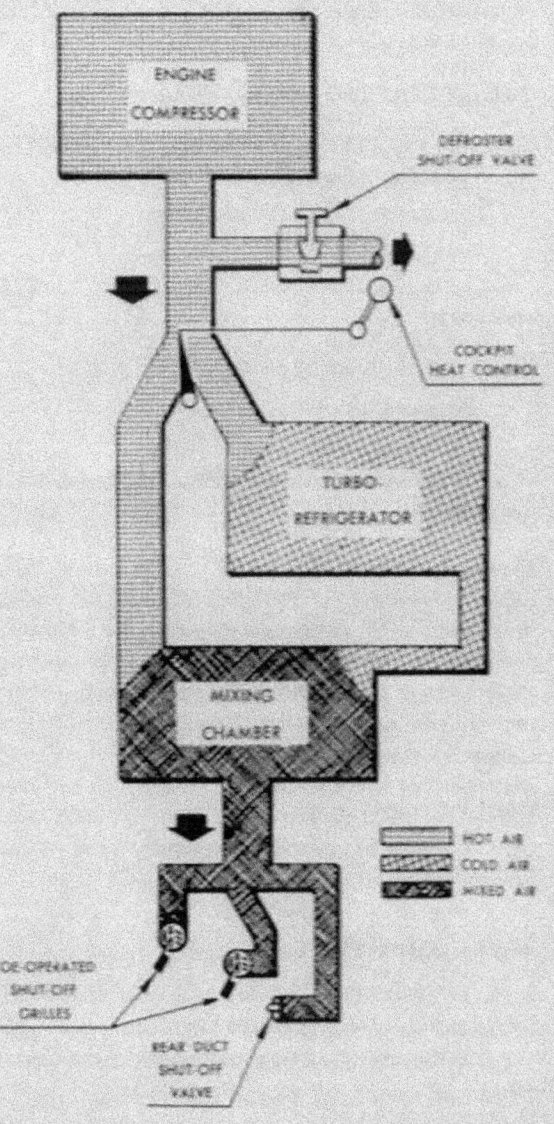

Figure 4-1 — Heating & Pressurization Diagram

4-9. **RELIEF AND DUMP VALVE.** The combination pressure relief, vacuum relief and dump valve unit operates automatically to relieve excessive cabin pressure in case of failure of the pressure regulator, and is operated manually as a cabin pressure dump valve. The toe operated dump valve control lever (figure 1-6, reference 38) is located at the floor forward of the control stick. The lever is pushed down to release cabin pressure when it is desired to heat or cool the cockpit without pressurization.

4-10. **PRESSURE REGULATOR.** Cockpit pressure differential is automatically maintained by the pressure regulator. From sea level to 8000 feet, the cockpit is unpressurized, between 8000 and 15,300 feet the cockpit is pressurized to maintain 8000 feet cabin altitude. Above 15,300 feet cockpit pressure is maintained at 2.75 psi above outside air pressure.

4-11. **OPERATING INSTRUCTIONS.**

4-12. **TO CONTROL COCKPIT TEMPERATURE** (without pressurization).

   a. Pressure shut-off valves—Open.

   b. Relief and dump valve—Open.

   c. Cockpit heat control—Adjust to obtain desired temperature.

4-13. **TO PRESSURIZE COCKPIT.**

   a. Pressure shut-off valves—Open.

   b. Relief and dump valve—Closed.

   c. Cockpit heat control—Adjust as desired to control temperature.

4-13A. **PRESSURIZED SUIT VALVE** — A "G" suit valve is installed on the left side of each cockpit at floor level. The valve receives air under pressure from the engine compressor and meters it to the pilot's pneumatic suit during positive "G" accelerations. At the high setting, suit pressurization begins at 1.5 "G" and increases at the rate of 1.4 psi per "G." At the low setting suit pressurization begins at 1.7 "G" and increases at the rate of 1 psi per "G." The suit is connected to the "G" valve by a quick disconnect fitting at the left side of the seat.

4-14. DEFROSTING.

4-15. **DESCRIPTION.** Hot air for defrosting the windshield and canopy is obtained from the engine compressor (see figure 4-1), and is distributed by a perforated tube.

4-16. **SHUT-OFF CONTROL.** The control knob in the defrosting tube forward of the gun sight controls the defrosting system.

4-17. **OPERATING INSTRUCTIONS.** To defrost, push the control knob in and turn it to engage the knob in the open position.

4-18. **AUXILIARY DEFROSTER.** An electrically operated windshield defroster is provided for cold weather operation or when the normal hot air system is insufficient for the weather conditions. The control switch is located on the right-hand shelf (figures 1-8 and 1-9, references 15 and 12).

4-19. OXYGEN SYSTEM.

4-20. **GENERAL.** A low pressure oxygen system consisting of four Type D-2 and one Type F-1 oxygen cylinders is installed in the airplane. The Type D-2 cylinders are located two in each wing, and the Type F-1 cylinder is located in the fuselage. The system may be refilled through a single filler valve which is located in the engine air intake duct on the left side of the fuselage. The oxygen pressure gages (2, fig. 1-6 and 41, fig. 1-7) and flow indicators (3, fig. 1-6 and 42, fig. 1-7) are located on the lower left side of the instrument panels in each cockpit. The Type A-14 pressure breathing diluter demand oxygen regulator (37, fig. 1-4 and 18, fig. 1-5) is located on the inboard side of the left console in both cockpits. Only a pressure breathing demand oxygen mask should be used.

4-21. **NORMAL OPERATION.** The diluter lever of the oxygen regulator should always be set at the "NORMAL OXYGEN" position except under emergency conditions. The pressure dial of the oxygen regulator should be set as follows:

   a. For cabin altitudes below 30,000 feet, leave dial at "NORMAL" position.

   b. For cabin altitudes between 30,000 feet and 40,000 feet, set the pressure dial at "SAFETY" position.

   c. For cabin altitudes above 40,000 feet, set the pressure dial to the cabin altitude.

4-22. **EMERGENCY OPERATION.** With symptoms of the onset of anoxia, set the diluter lever to "100% OXYGEN." If the oxygen regulator becomes inoperative, pull the cord of the H-2 emergency oxygen cylinder. If smoke or fuel fumes should enter the cabin, proceed as follows:

   a. Set oxygen regulator diluter lever to "100% OXYGEN" position.

   b. Set pressure dial of oxygen regulator as required by cabin altitude.

CREW OXYGEN DURATION — HOURS

CABIN ALTITUDE — FEET	GAGE PRESSURE — P.S.I.							BELOW 100
	400	350	300	250	200	150	100	
40,000	4.2 / 4.2	3.6 / 3.6	3.0 / 3.0	2.4 / 2.4	1.8 / 1.8	1.2 / 1.2	0.6 / 0.6	EMERGENCY DESCEND TO ALTITUDE NOT REQUIRING OXYGEN
35,000	4.2 / 4.2	3.6 / 3.6	3.0 / 3.0	2.4 / 2.4	1.8 / 1.8	1.2 / 1.2	0.6 / 0.6	
30,000	3.2 / 3.2	2.7 / 2.7	2.2 / 2.2	1.8 / 1.8	1.4 / 1.4	0.9 / 0.9	0.5 / 0.5	
25,000	2.5 / 3.0	2.2 / 2.6	1.8 / 2.1	1.4 / 1.7	1.1 / 1.3	0.7 / 0.9	0.4 / 0.4	
20,000	2.0 / 3.4	1.7 / 2.9	1.5 / 2.4	1.2 / 1.9	0.9 / 1.4	0.6 / 1.0	0.3 / 0.5	
15,000	1.6 / 4.1	1.4 / 3.5	1.1 / 2.9	0.9 / 2.3	0.7 / 1.8	0.5 / 1.2	0.2 / 0.6	
10,000	1.4 / 5.4	1.2 / 4.7	1.0 / 3.9	0.8 / 3.1	0.6 / 2.3	0.4 / 1.6	0.2 / 0.8	

Black figures indicate diluter lever "NORMAL."

Red figures indicate diluter lever "100%."

Cylinders: 4 Type D-2, 1 Type F-1.

Crew: 2.

4–23. ARMAMENT.

4–24. GUNNERY EQUIPMENT. Two .50 caliber guns with ammunition boxes carrying up to 300 rounds each are located in the nose armament compartment. A gun camera mounted in the lip of the right engine intake duct, operates with the guns or separately. The gunsight is mounted above the instrument panel.

4-25. CONTROLS. The gun camera switch (figure 1-4, reference 29) has three positions, "SIGHT AND CAMERA," "OFF," and "GUNS." When the switch is in the "SIGHT AND CAMERA" position only the gunsight and the gun camera are operable. When the switch is in the "GUNS" position, the gunsight, gun camera and the .50 caliber machine guns are operable. The guns and camera are operated by squeezing the control stick trigger.

4–26. BOMBING EQUIPMENT. The following items may be carried on the wing tip bomb shackles in place of the droppable fuel tanks.

100 pound practice bombs, type M38A2

500 pound general purpose bombs, type AN-M64A1, only when bomb fins are modified and M2 fin lock nuts are used.

1000 pound general purpose bombs, type AN-M65A1, only when bomb fins are modified and M2 fin lock nuts are used.

MK47 practice bombs only when bomb racks and aircraft are modified.

4–26A. BOMB CONTROLS. Bomb controls consist of the bomb arming and selector switches (figure 1-4, reference 32), in the front cockpit and bomb arming and release master switches in the rear cockpit (figure 1-5, references 22 and 23). A bomb release button is incorporated in the control stick grip, in each cockpit. A bomb salvo switch (figures 1-6 and 1-7, references 40 and 38) is provided in each cockpit to permit release of the bombs simultaneously (without presetting the master or arming and selector switches) in an emergency.

Revised 15 November 1950

Section IV
Paragraphs 4-26B to 4-31A

4-26B. OPERATION OF BOMB CONTROLS. The bomb arming and release master switches in the rear cockpit must be in the "ARM" position before the arming and selector switches in the front cockpit will function. To arm the bombs, place the bomb arming switch (front cockpit) in the "TAIL" or "NOSE AND TAIL" position. To drop the bombs individually, place the bomb selector switch in the "TRAIN" position and press the bomb release button on the control stick grip. The left bomb will drop first, the right bomb will drop the next time the bomb release button is pressed. To drop both bombs simultaneously, place the bomb selector switch in the "ALL" position and press the bomb release button.

4-26C. CHEMICAL TANK CONTROLS. The chemical tank circuit is connected with the bomb release circuit through the master switches in the rear cockpit. The chemical tank selector switch is located in the front cockpit (figure 1-4, reference 31). To operate the chemical tanks individually, place the chemical tank switch in the "LEFT" or "RIGHT" position and press the bomb release button on either control stick. To operate the tanks simultaneously, place the bomb arming switch (front cockpit) in the "NOSE AND TAIL" position. To release the chemical tanks, follow the same procedure given for bombs or drop tanks.

4-26D. TOW TARGETS.

4-26E. Either the fuselage type installation or the jato latch type installation may be used. With either type of installation a banner type A-6B target is used.

4-26F. RELEASE OF JATO LATCH TOW TARGETS.

a. To accomplish release of the target when attached to the jato latch, the tow plane should be flown at minimum safe flying speed. This procedure reduces the target drag load and facilitates operation of the manual release.

b. In case of failure to release the target, a landing should be made at an adequate distance from the end of the runway in order that the target may clear all obstacles short of the runway. The glide angle should be planned accordingly with an increase in prescribed approach and landing speed of approximately 10 miles per hour IAS.

4-27. LIGHTING.

4-28. COCKPIT LIGHTING. Two fluorescent and one incandescent focusing cockpit light are located on each shelf, in each cockpit. Each fluorescent light is controlled by an integral rheostat.

4-28A. Airplanes serials 50-402 and subsequent are provided with red floodlighting. Pairs of red floodlights, controlled by a switch type rheostat on each left side panel, provide additional lighting for the instrument panel and left and right side panels in each cockpit.

4-29. LANDING AND TAXI LIGHT. The landing light switch (figures 1-4 and 1-5, references 13 and 8) is wired so that the center position is "OFF." The "TAXI" position supplies power to one of the two lights mounted on the nose landing gear strut and the "LANDING" position supplies power to both lights.

CAUTION

The landing lights must not be left on for longer than five minutes when the airplane is on the ground.

4-29A. EXTERIOR LIGHTS. Navigation, fuselage signal lights, and the flasher-coder switches are located in the front cockpit only.

4-29B. NAVIGATION LIGHTS. The navigation lights are controlled by the "DIM-BRIGHT" switch and the "STEADY-OFF-FLASH" switch (figure 1-8, reference 12).

4-29C. FUSELAGE LIGHTS. The six watt fuselage lights, located on the top and bottom of the fuselage, are controlled by a switch (figure 1-8, reference 10A) having three positions, "DIM," "OFF," and "BRIGHT."

4-29D. SIGNAL LIGHTS. The 100 watt signal lights are mounted on the top and bottom of the fuselage adjacent to the fuselage lights. The lights may be used to flash in Morse code the letters selected on the code sequence selector (figure 1-8, reference 16A). The signal lights are controlled by an "ON-OFF" switch. A signal light indicator is provided to indicate when the signal lights are turned on and blink with the coded signals.

4-30. COMMUNICATIONS EQUIPMENT.

4-31. AN/AIC-2A INTERPHONE. The interphone amplifier is on whenever the battery switch is on. A microphone selector switch and indicator light is located in each cockpit in early airplanes. Place the switch in "RADIO" position before transmitting, and in the "INTERPHONE" position at all other times. The indicator light is off when the microphone is connected to the AN/ARC-3 radio.

4-31A. USAF COMBAT INTERPHONE SYSTEM. Later airplanes are equipped with the USAF Combat Interphone Components system which includes a mixer amplifier and a control panel (1, figure 1-4; and 7A, figure 1-9) for each cockpit. This combination enables

RESTRICTED
AN 01-75FJC-1

Section IV
Paragraphs 4-32 to 4-37

TABLE OF ELECTRONIC EQUIPMENT

Type	Use	Range	Illustration of Controls
AN/AIC-2A Interphone	Intercrew communication		Figures 1-4 and 1-5
USAF Combat Interphone (later airplanes)	Intercrew communication		Figures 1-4 and 1-9
AN/ARC-3 Tranceiver	VHF Short Range Two-Way Communication	Line of Sight	Figures 1-4 and 1-9
AN/ARN-6 ADF Radio Compass Receiver	Navigation—Aural Reception Bearing—Homing	500 Miles	Figures 1-6 and 1-7
RC-193-A Marker Beacon Receiver	Navigation Aid		Figures 1-6 and 1-7
AN/APX-6	IFF		

the operator to listen to any one or all audio channels simultaneously. The control panel includes a volume control, range filter switch, a rotary channel selector switch and four individual channel selector toggles. The system is in operation whenever the airplane battery switch is on. The system is operated by selecting the facility desired with the rotary selector switch. To transmit, press the microphone button on the throttle and speak into the microphone.

4-32. AN/ARC-3 VHF TRANCEIVER. The AN/ARC-3 command set is a short range, multi channel, two-way voice or MCW, transmitter-receiver. The control panels located in each cockpit (figures 1-4 and 1-9, references 6 and 7) include a volume control, eight channel selector buttons, an "OFF" button and lock, and a tone control button. A control switch and indicator light are also provided in each cockpit (figures 1-4 and 1-9, references 5 and 8) to switch control from one cockpit to the other.

4-32A. OPERATION—AN/ARC-3 TRANCEIVER.

a. If the AN/ARC-3 control indicator light is not on, push the control switch until the light comes on.

b. To turn the set on push the desired channel selector button. Allow about one minute for the set to warm up.

c. To transmit, first push the "MIC SELECTOR" switch to the "RADIO" position until the indicator light goes out, then press the microphone button on the throttle and speak.

d. To transmit code use the tone control button as a key.

e. To turn the set off, push the "OFF" button and the adjacent lock button.

4-33. AN/ARN-6 RADIO COMPASS—OPERATION. Turn the set on by pushing the control switch and then turning the function switch (figures 1-6 and 1-7, references 37 and 32) to "COMP," "ANT," or "LOOP." Turn the set off by turning the function switch to "OFF."

CAUTION

Erratic operation of the compass can be expected when the airplane is in a bank of 45 degrees or more as manifested by any one of the following:

a. The compass indicator may slow down so that there is considerable lag between the beaming indication and the airplane heading.

b. The indicator may stop altogether or move in jerks.

c. The indicator may start moving in the opposite direction from the turn, later change direction but still lag the airplane heading considerably.

d. The indicator may oscillate violently. Since the erratic operation occurs only during a turn it will not seriously affect the use of the radio compass.

4-34. RC-193-A or AN/ARN-12—MARKER BEACON RECEIVER. The marker beacon receiver is on whenever the battery switch is on. The indicator light is located on the instrument panel.

4-35. DE-ICING EQUIPMENT.

4-36. FUEL FILTER DE-ICING. The filter de-icing system utilizes components of the fluid injection system; therefore, if the airplane is serviced for filter de-icing, fluid injection will not be available and vice versa. The choice of fluid injection or fuel filter de-icing will depend on the ambient air temperature. Filter icing may occur when the fuel temperature reaches +15°F or lower, and use of fluid injection is prohibited at +32°F or lower ambient ground temperatures.

4-37. FUEL FILTER DE-ICING SYSTEM. The fuel filter de-icing system utilizes the right-hand fluid injec-

Section IV
Paragraph 4-37

tion tank and pump. In addition, a warning light and de-icing switch (figures 1-4 and 1-5, references 7 and 9) are located on the left-hand auxiliary panel adjacent to the throttle. The warning light is operated by a differential pressure switch which senses the fuel pressure drop across the low pressure fuel filter. If the filter pressure drop reaches approximately 2 psi, the warning light comes on, indicating the possibility of icing. When the airplane is serviced for filter de-icing, the right-hand fluid injection tank is filled with 100% AN-A-18 alcohol and the tank is connected through the fluid injection pump and a solenoid shut-off valve to the low pressure fuel filter. Holding the de-icer switch in the "ON" position opens the solenoid valve and pumps alcohol into the filter. If the filter is iced, the alcohol will dissolve the ice accumulation, reducing the pressure drop, and the warning light will go out.

WARNING

If the filter icing warning light comes on, hold the de-icing switch in the "ON" position until the warning light goes out. If the warning light does not go out after holding the de-icing switch on for from 20 to 30 seconds, the filter may be clogged with dirt and should therefore be inspected as soon as possible.

APPENDIX I
FLIGHT OPERATING DATA

A-1. FLIGHT OPERATION INSTRUCTION CHARTS.

A-2. The purpose of the Flight Operation Instruction Chart is to show the range for the fuel remaining in the airplane and the procedure required to obtain this range. The main variables affecting range have been incorporated in an effort to give the most usable and most accurate information consistent with simplicity.

A-3. The chart may be used at any point in flight or for preflight planning. The initial conditions are the actual altitude of the airplane and the fuel remaining on board. In the Flight Operation Instruction Chart, the five main colums across the top are initial altitude conditions. On line opposite fuel quantities, ranges are shown for each initial altitude. In general, two range values are given for each altitude and fuel quantity, one for level flight at that altitude, and one for the maximum range obtainable by climbing to a higher altitude. Distances covered in let-down are included, and for range figures indicating a cruise at higher altitude, climb distance is included. Ranges quoted allow no fuel reserve for landing after descent to sea level.

A-4. FUEL QUANTITIES. Fuel quantities tabulated on the chart represent fuel that is available for cruising and landing. Allowances must be made for extra items such as combat and endurance reserves. Additional allowances must be made for evaporation losses when using gasoline and JP-3 fuels and for fuel "slugging" losses when using JP-3 fuel under adverse conditions. During fuel "slugging" large quantities of solid fuel are carried overboard through the vent system by violent foaming of the fuel. The fuel quantities to allow for these losses cannot be simply presented as they vary from zero to considerable amounts depending upon atmospheric temperature, fuel temperature at take-off; individual fuel shipments, the length of time since the fuel was refined (amount of weathering) and the rate of change of altitude during flight.

A-5. WIND. Under different wind conditions ranges are varied by the effect of wind on ground speed. Let-down distances are affected for the same reason. Recommended airspeed to obtain long range may also change with different headwinds in order to maintain the most favorable miles-per-gallon ratio. The lower half of the Flight Operation Instruction Chart contains operating instructions for different wind conditions. These cruising data are presented for the same five altitudes that head the upper half of the chart.

A-6. Since the wind may be from any direction with respect to the airplane course, some question may arise as to the method of handling winds other than straight headwinds or tailwinds. For purposes of cruise control, all winds may be expressed as effective winds. This reduces the wind to one which would have the same effect on the airplane's ground speed if it were a straight head or tailwind. In other words, it is the component of wind in the direction of the airplane heading. For example, a 100 mph wind at 45 degrees to the course will be an effective headwind of about 70 mph. For an airplane whose still air cruising speed is 400 mph, the ground speed along the course will be about 330 mph.

A-7. TAKE-OFF CHART.

A-8. The Take-off Chart (figure A-4) lists take-off distance for various pressure altitudes and air temperatures.

A-9. Set airplane altimeter to 29.92 and read pressure altitude. With air temperature in degrees Fahrenheit as obtained from the field weather station and pressure altitude, enter chart and determine required take-off distances.

A-10. The Take-off and Landing Chart is based on flight test data that were obtained by following the procedures outlined in Section II.

A-11. USE OF THE FLIGHT OPERATION INSTRUCTION CHARTS.

A-12. To use the charts in flight, the pilot refers to the upper half, and under the present altitude column reads range opposite fuel quantity. For cruising at that altitude the operating instructions are listed directly below. Entering on the line according to effective wind, read the range factor, calibrated airspeed, and let-down distances. Multiplying still air range by the range factor results in ground miles that can be flown. Approximate values of cruising rpm, gallons per hour, and ground miles per hour are given for reference.

A-13. If it is desirable to increase range enter the same altitude column as before. Under the second and third subheadings are shown the optimum altitude to which a climb should be made to obtain best range, and the range at that optimum altitude. To obtain this range climb immediately (according to the recommended climb procedure) to the altitude shown. For cruising instructions refer to the lower half of the chart in the column according to the new altitude. Calculation of range in a wind and cruising procedure are as described above for the level flight cruise. Note that at any time during the flight, the pilot may refer to the chart with actual conditions of altitude and fuel to obtain range remaining in the same manner as previously discussed.

A-14. EXAMPLES OF USE OF CHARTS.

A-15. Maximum range on internal fuel (353 gallons) at 35,000 feet altitude against an 80 mph headwind. Take-off weight 11,350 lb. (Water tanks empty.)

a. From the climb chart it is seen that the take-off and climb to altitude will use 113 gallons of fuel. The still air range covered in climb will be about 73 miles. The fuel remaining at 35,000 feet will be 240 gallons (353-113).

b. By referring to the 35,000 foot section of the Flight Operation Instruction Chart (figure A-7, sheet 2) opposite 240 gallons, it can be seen that 620 additional still air miles can be flown, including allowance for let-down. The total still air range is then 73 plus 620 or 693 miles.

c. In the lower half of the chart it is seen that the range factor for an 80 mph headwind is .8. Multiplying the still air range by this factor gives about 555 miles actual range.

d. Cruising at 35,000 feet with a headwind of 80 mph, according to the lower half of the Flight Operation Instruction Chart, is at 266 mph CAS and the let-down is begun 64 miles from the destination.

e. Fuel reserve allowances (for landing, loitering, etc.) should be subtracted from fuel capacity before obtaining data from charts.

A-16. Illustration of the use of the chart in flight. The airplane is at 5,000 feet altitude with 330 gallons of fuel and distance to destination is 500 miles. A 50 gallon fuel reserve is desired for landing.

a. Reference to the 5,000 feet column of the Flight Operation Instruction Chart opposite 280 gallons (330-50) shows that by cruising at 5,000 feet the range will be only 275 miles. By climbing to 40,000 feet a flight of 615 miles can be made. In order to fly 500 miles it is evident that it is necessary to climb and cruise at an altitude higher than 5,000 feet, but not necessarily as high as 40,000 feet. A linear interpolation (which in all cases will be close to the actual values) between the difference in range $(615 - 275 = 340)$ and altitude $(40,000 - 5,000 \text{ feet} = 35,000)$ provides a quick guess that for the 225 additional miles of range needed $(500 - 275 = 225)$ an increase of at least 23,000 feet of altitude will be necessary (or a minimum cruising altitude of 28,000 feet). The 500 miles are available by climbing and cruising at 30,000 feet.

b. For purpose of checking the estimate detailed calculations are shown. A distance of 49 miles will be covered with an expenditure of 66 gallons of fuel in climbing from 5,000 feet to 30,000 feet. This means that there are only $500 - 49$ or 451 miles to go from that point and 214 gallons are available. With these as the initial conditions enter the Flight Operation Instruction Chart in the 30,000 foot column. The distance which can be flown at 30,000 feet opposite 200 gallons is 450 miles. This shows that a climb to 30,000 feet will provide sufficient range to reach destination and leave a 50 gallon landing reserve. (Cruising speed at 30,000 feet is 278 mph CAS.)

A-17. ESCORT MISSION.

A-18. It is desired to escort bombers at 25,000 feet, tip tanks to be carried and dropped when empty. Fifteen minutes combat at 100% rpm at 25,000 feet to be included. How far can the bombers be escorted? A 50 gallon reserve is desired for safe landing.

A-19. The take-off fuel will be 683 gallons. The combat allowance chart indicates that 105 gallons (15 minutes at 7 gallons per minute) will be required for combat.

a. The climb chart shows 111 gallons will be used and 51 miles will be covered in climb to altitude. (Fuel for take-off included in the 111 gallons.)

b. After 25,000 feet is reached $683 - 111 = 572$ gallons will be available for level flight, combat, descent, and 50 gallon landing reserve. Subtracting the 105 gallon allowance for combat and 50 gallons for reserve leaves 417 gallons. Reference to figure A-7A, sheet 2 shows that at 25,000 feet 789 miles can be flown with 417 gallons of fuel. With the 51 miles covered in climb $789 + 51 = 840$ miles can be covered.

The bombers can be escorted $\frac{840}{2} = 420$ statute miles.

c. The operating instructions on the lower half of figure A-7A, sheet 2 show that at 25,000 feet 283 mph CAS is required and the engine speed is approximately 89% rpm. Figure A-7, sheet 2 (to be used after tip tanks are dropped) shows at 25,000 feet 290 mph CAS is required and engine speed is approximately 87% rpm.

d. Reference to the upper half of figure A-7, sheet 2 shows that at 25,000 feet approximately 213 gallons will be required for the return trip (420 miles). If a climb is made to 40,000 feet for the return trip, 552 miles can be covered with 213 gallons. This would provide a reserve of approximately 112 miles $(552 - 420)$ in addition to the 50 gallon reserve.

A-20. Deleted.

A-21. MAXIMUM FERRY RANGE.

A-22. Take-off fuel with tip tanks = 683 gallons. Tip tanks to be carried all the way. Allow a 50 gallon reserve for landing.

A-23. Reference to figure A-7B shows that the optimum altitudes for any fuel quantity over 350 gallons is 40,000 feet.

The climb chart shows that 176 gallons and 145 miles will be covered in taxi, take-off and climb to 40,000 feet.

b. After 40,000 feet is reached 683 — 176 — 50 = 457 gallons will be available for level flight and let-down.

c. For 457 gallons at 40,000 feet about 1,151 miles are available.

d. With the 145 miles covered in climb a total flight of 1,151 + 145 = 1,296 miles can be made.

A-24. Reference to figure A-7A, sheet 2 (tip tanks dropped when empty) shows that at 40,000 feet, 457 gallons will permit a flight of 1,237 miles. With the 145 miles covered in climb the total range with a 50 gallon landing reserve is 1,237 + 145 = 1,382 miles.

COMBAT ALLOWANCE CHART DROP TANKS OFF		
AT ALTITUDE (FT.)	FUEL REQUIRED U.S. GAL. PER MIN.	
	96% RPM	100% RPM
40,000	3	4
35,000	4	5
30,000	5	6
25,000	5	7
20,000	6	8
15,000	7	9
10,000	9	11
5,000	10	12
S.L.	12	14

Figure A-1 — Combat Allowance Chart

Appendix I

RESTRICTED
AN 01-75FJC-1

AIRPLANE
MODEL
T-33A

ENGINE
MODELS
J33-A-23
J33-A-35

FUEL SPEC MIL-F-5616 (AN-F-32) (GRADE JP-1)
MIL-F-5624 (AN-F-58) (GRADE JP-3)
OR ALTERNATE:
GASOLINE MIL-F-5572 (AN-F-48) (GRADE 100/130)

OIL SPEC AN O-9 (GRADE 1010) OR AF 3606

OPERATING CONDITION	R. P. M.	TIME LIMIT
TAKE-OFF	100%	30 MINUTES
MILITARY	100%	30 MINUTES
MAX. CONT.	96%	NO LIMIT

Figure A-2 — Engine Operating Limits

INSTR. I.A.S.	CORRECT I.A.S.				
	S.L.	10,000	20,000	30,000	40,000
100	100				
125	125				
150	150	150	149	148	146
175	175	174	174	172	170
200	200	199	198	196	193
225	224	223	221	219	215
250	249	247	245	242	237
275	273	271	268	264	257
300	298	295	292	286	
325	322	319	315	308	
350	346	342	337	329	
375	371	366	360		
400	395	389	382		
425	419	413	404		
450	444	436			
475	468	460			
500	493	483			
525	517	506			
550	542				
575	567				

Figure A-3 — Airspeed Correction Table

RESTRICTED
AN 01-75FJC-1

TAKE-OFF DISTANCES
FEET

AIRPLANE MODEL T-33A **ENGINE MODEL J33-A-23 / J33-A-35**

70% FLAPS – HARD SURFACE RUNWAY – NO WIND

CONFIGURATION AND GROSS WEIGHT	PRESS. ALT. FT.	40°F GROUND ROLL	40°F CLEAR 50'	60°F GROUND ROLL	60°F CLEAR 50'	80°F GROUND ROLL	80°F CLEAR 50'	100°F GROUND ROLL	100°F CLEAR 50'	120°F GROUND ROLL	120°F CLEAR 50'
CLEAN 11,750 LB WITHOUT WATER INJECTION	5000	2550	3725	2975	4325	3425	5000	3950	5800	4500	6675
	4000	2350	3425	2725	3950	3125	4575	3600	5275	4075	6025
	3000	2150	3150	2475	3625	2875	4200	3275	4800	3725	5500
	2000	1975	2875	2275	3325	2625	3850	3000	4400	3425	5025
	1000	1825	2650	2100	3050	2400	3525	2750	4025	3125	4600
	S.L.	1675	2450	1925	2800	2200	3225	2525	3675	2850	4200
CLEAN 11,750 LB WITH WATER INJECTION	5000	2150	3225	2490	3750	2875	4325	3325	4975	3775	5700
	4000	1975	2975	2300	3450	2650	3975	3025	4550	3450	5200
	3000	1825	2750	2100	3175	2425	3650	2775	4150	3150	4750
	2000	1675	2525	1925	2900	2225	3350	2525	3800	2875	4350
	1000	1525	2325	1775	2675	2025	3050	2300	3475	2625	3950
	S.L.	1425	2150	1625	2450	1850	2800	2125	3000	2400	3625
TIP TANKS 14,250 LB WITHOUT WATER INJECTION	5000	3600	5475	4125	6300	4750	7300	5525	8475	6325	9750
	4000	3325	5025	3800	5800	4375	6700	5050	7750	5800	8875
	3000	3050	4625	3475	5325	4025	6150	4600	7075	5300	8100
	2000	2775	4250	3200	4875	3675	5625	4200	6475	4825	7400
	1000	2550	3900	2925	4475	3375	5175	3850	5925	4400	6750
	S.L.	2350	3600	2700	4125	3100	4750	3550	5425	4025	6175
TIP TANKS 14,250 LB WITH WATER INJECTION	5000	2950	4700	3400	5450	3900	6275	4500	7275	5125	8375
	4000	2750	4325	3150	5000	3600	5775	4150	6675	4700	7650
	3000	2525	3975	2900	4600	3300	5300	3800	6100	4300	6950
	2000	2300	3650	2650	4200	3025	4850	3475	5575	3925	6325
	1000	2125	3350	2425	3850	2775	4425	3175	5075	3575	5775
	S.L.	1950	3075	2225	3550	2550	4050	2900	4625	3300	5300

NOTES: 1. FOR HEADWINDS—DECREASE T.O. DISTANCE 1% FOR EACH 1.0 MPH HEADWIND VELOCITY.
2. DATA WERE OBTAINED WITH J33-A-23 ENGINES, AND ARE CONSERVATIVE FOR AIRPLANES WITH J33-A-35 ENGINES.

DATA AS OF: 1-1-49
BASED ON: FLIGHT TEST (See Note 2)

BASED ON MIL-F-5516 (AN-F-32) GRADE JP-1 FUEL
RED FIGURES HAVE NOT BEEN FLIGHT CHECKED

Figure A-4 — Take-off Distances

Revised 15 November 1950

Appendix I

RESTRICTED
AN 01-75FJC-1

LANDING DISTANCE
FEET

AIRPLANE MODEL
T-33A

ENGINE MODEL
J33-A-23
J33-A-35

GROSS WEIGHT LB.	BEST CAS APPROACH		100% FLAPS — HARD SURFACE — NO WIND							
	POWER OFF	POWER ON	AT SEA LEVEL		AT 2000'		AT 4000'		AT 6000'	
	MPH	MPH	GROUND ROLL	CLEAR 50'	GROUND ROLL	CLEAR 50'	GROUND ROLL	CLEAR 50'	GROUND ROLL	CLEAR 50'
10,000	125	125	2350	3275	2500	3450	2650	3625	2800	2900
12,500	140	140	2950	3975	3125	4190	3300	4400	3500	4650

NOTES:

LEGEND
CAS — CALIBRATED AIRSPEED
MPH — MILES PER HOUR

DATA AS OF: 1-1-49
BASED ON: FLIGHT TEST

BASED ON MIL-F-5516 (AN-F-32) GRADE JP-1 FUEL
RED FIGURES HAVE NOT BEEN FLIGHT CHECKED

Figure A-5 Landing Distances
RESTRICTED
Revised 15 November 1950

RESTRICTED
AN 01-75FJC-1

Appendix 1

| AIRPLANE MODEL T-33A | CLIMB CHART STANDARD DAY (59° F AT SEA LEVEL) | ENGINE MODEL J33-A-23 |

Rate of Climb and Rate of Descent given in feet per minute.

100% RPM					PRESSURE ALTITUDE FEET		100% RPM			
	APPROXIMATE						APPROXIMATE			
RATE OF CLIMB	FROM SEA LEVEL			CAS MPH		CAS MPH	FROM SEA LEVEL			RATE OF CLIMB
	DISTANCE	TIME	FUEL				FUEL	TIME	DISTANCE	
	WITH DROP TANKS 14,250 LBS. GROSS WEIGHT				AIRPLANE CONFIGURATION & GROSS WEIGHT		CLEAN CONFIGURATION 11,750 LBS. GROSS WEIGHT			
4300	0	0	30 (1)	310	SEA LEVEL	310	20 (1)	0	0	5600
3800	8	1.2	45	300	5,000'	300	32	1	5	4800
3300	16	2.6	61	290	10,000'	290	44	2.2	12	4150
2800	25	4.3	77	280	15,000'	280	57	3.5	19	3550
2350	37	6.3	93	270	20,000'	270	70	5	28	3000
1950	51	8.6	111	260	25,000'	260	84	6.8	40	2500
1550	70	11.5	128	250	30,000'	250	98	9.2	54	2000
1150	97	15.3	148	240	35,000'	240	113	12.2	73	1550
350	145	21.9	176	230	40,000'	230	131	16.5	106	700
	WITH DROP TANKS 11,250 LBS. GROSS WEIGHT				AIRPLANE CONFIGURATION & GROSS WEIGHT		CLEAN CONFIGURATION 9,750 LBS. GROSS WEIGHT			
5400	0	0	30 (1)	310	SEA LEVEL	310	20 (1)	0	0	6600
4800	5	1.2	42	300	5,000'	300	30	0.8	4	5750
4250	11	2.3	54	290	10,000'	290	40	1.8	9	5000
3700	19	3.5	66	280	15,000'	280	51	2.8	15	4350
3200	27	5.0	78	270	20,000'	270	61	4.0	23	3750
2700	37	6.6	90	260	25,000'	260	72	5.4	32	3150
2250	51	8.6	102	250	30,000'	250	83	7.2	43	2650
1750	70	11.3	117	240	35,000'	240	94	9.5	57	2100
900	96	14.6	132	230	40,000'	230	106	12.3	80	1150

NOTES: (1) Taxi and take-off allowance.
(2) To correct Rate of Climb values for air temperature different from standard day temperature subtract 35 ft. per min. from the Rate of Climb for every °F above standard day temperature for both clean configuration and drop tank configuration.
(3) Data were obtained from J33-A-23 Engines and are conservative for J33-A-35 Engines.

NOTES: (1) Descend at .6 mach number
(2) Use dive flaps down to 35,000 if idle rpm is too great to allow descent at .6 mach number

DESCENT CHART
STANDARD DAY

AIRPLANE CONFIGURATION & GROSS WEIGHT WITH DROP TANKS 9600 LBS. GROSS WEIGHT					PRESSURE ALTITUDE FEET		AIRPLANE CONFIGURATION & GROSS WEIGHT CLEAN CONFIGURATION 9,100 LBS. GROSS WEIGHT			
	APPROXIMATE						APPROXIMATE			
RATE OF DESCENT	TO SEA LEVEL			CAS MPH		CAS MPH	TO SEA LEVEL			RATE OF DESCENT
	DISTANCE	TIME	FUEL				FUEL	TIME	DISTANCE	
1200	85	12.5	19	200	40,000	200	25	15.2	98	1000
1700	63	8.7	14	230	35,000	230	19	9.6	70	1500
2400	46	6.4	11	260	30,000	260	15	7	51	2150
3200	34	4.7	8	285	25,000	285	11	5.2	37	2850
4100	24	3.3	6	315	20,000	315	8	3.7	27	3700
5150	16	2.2	4	350	15,000	350	6	2.5	18	4650
6300	9	1.4	3	385	10,000	385	3	1.5	11	5750
7550	4	0.6	1	420	5,000	420	2	0.8	5	6850
8900	0	0	0	455	SEA LEVEL	455	0	0	0	8050

MULTIPLY STATUTE UNITS BY .87 FOR CONVERSION TO NAUTICAL UNITS
DATA AS OF: 7-1-49
BASED ON: FLIGHT TEST
BASED ON: AN-F-32 (JP-1) FUEL
RED FIGURES HAVE NOT BEEN FLIGHT CHECKED

LEGEND
FUEL – U.S. GALLONS
DISTANCE – STATUTE MILES
TIME – MINUTES
RATE OF CLIMB – FEET PER MINUTE
CAS – CALIBRATED AIRSPEED
MPH – STATUTE MILES PER HOUR

Figure A-6 — Climb and Descent Chart

Revised 15 November 1950

Appendix I

RESTRICTED
AN 01-75FJC-1

FLIGHT OPERATION INSTRUCTION CHART

AIRPLANE MODEL(S)		EXTERNAL LOAD ITEMS
T-33A		NONE

ENGINE(S)	CHART WEIGHT LIMITS 11,800 TO 8200 POUNDS	NUMBER OF ENGINES OPERATING: ONE
J33-A-23, J33-A-35		

INSTRUCTIONS FOR USING CHART. (A) IN FLIGHT — Select figure in fuel column equal to or less than fuel available for cruise (fuel on board minus allowances for reserve, combat, navigational errors, formation flight, etc.). Move horizontally right or left to section containing its present altitude and read total range available (no wind) by cruising at that altitude or by climbing to another altitude of maximum range. For a flight at initial altitude, operating instructions are given directly below. For a flight at higher altitude, climb immediately to desired altitude and read cruising instructions in appropriate cruising altitude section. (B) FLIGHT PLANNING — From initial fuel on board subtract fuel required for take-off and climb to desired cruising altitude and all other necessary allowances. Then use chart as for IN FLIGHT above, adding initial climb distances to range values.

NOTES: Ranges shown at optimum altitudes are maximum. In order to obtain maximum range on flights requiring more than one chart (due to external configuration or gross weight change), it is necessary to observe the optimum cruising altitude on each chart; i.e., when changing charts a climb may be required to obtain a maximum range. All range values include allowances for descent distance and fuel. Climb distance and fuel are included where climbs are indicated.

DATA BELOW CONTAIN NO FUEL RESERVE FOR LANDING

LOW ALTITUDE

IF YOU ARE AT S.L.			FUEL U.S. GAL.	IF YOU ARE AT 5000'			FUEL U.S. GAL.	IF YOU ARE AT 10,000'			FUEL U.S. GAL.	IF YOU ARE AT 15,000'			FUEL U.S. GAL.	IF YOU ARE AT 20,000'		
RANGE IN AIRMILES		Let Down Dist.		RANGE IN AIRMILES		Let Down Dist.		RANGE IN AIRMILES		Let Down Dist.		RANGE IN AIRMILES		Let Down Dist.		RANGE IN AIRMILES		Let Down Dist.
BY CRUISING AT S.L.	BY CRUISING AT OPT. ALT. 1000 FT.			BY CRUISING AT 5000'	BY CRUISING AT OPT. ALT. 1000 FT.			BY CRUISING AT 10,000'	BY CRUISING AT OPT. ALT. 1000 FT.			BY CRUISING AT 15,000'	BY CRUISING AT OPT. ALT. 1000 FT.			BY CRUISING AT 20,000'	BY CRUISING AT OPT. ALT. 1000 FT.	
285	775	40	350	345	800	40	350	420	840	40	350	495	865	40	350	575	910	40
255	695	40	320	315	720	40	320	380	760	40	320	450	785	40	320	530	825	40
225	590	40	280	275	615	40	280	335	650	40	280	395	675	40	280	460	715	40
190	480	40	240	240	505	40	240	290	540	40	240	340	565	40	240	400	605	40
160	375	40	200	200	400	40	200	240	430	40	200	285	460	40	200	335	495	40
130	270	40	160	160	295	40	160	195	320	40	160	230	350	40	160	270	380	40
95	160	25	120	102	190	35	120	150	215	40	120	180	245	40	120	205	270	40
65	—	—	80	80	100	25	80	100	115	25	80	120	135	25	80	140	155	30
30	—	—	40	40	—	—	40	50	—	—	40	65	—	—	40	80	—	—

(RANGE FIGURES INCLUDE ALLOWANCE FOR PRESCRIBED CLIMB AND DESCENT TO SEA LEVEL)

	CRUISING AT S.L.					EFFEC-TIVE WIND MPH	CRUISING AT 5000'					EFFEC-TIVE WIND MPH	CRUISING AT 10,000'					CRUISING AT 15,000'					CRUISING AT 20,000'				
			APPROXIMATE						APPROXIMATE						APPROXIMATE					APPROXIMATE					APPROXIMATE		
CAS	% RPM	GPH	G.S.	RANGE FACTOR	Let Down Dist.		CAS	% RPM	GPH	G.S.	RANGE FACTOR	Let Down Dist.		CAS	% RPM	GPH	G.S.	RANGE FACTOR	Let Down Dist.	CAS	% RPM	GPH	G.S.	RANGE FACTOR	Let Down Dist.	CAS	% RPM	GPH	G.S.	RANGE FACTOR	Let Down Dist.
417	88	530	377	.9	0	120 HW	378	86	430	366	.9	5	120 HW	354	85	355	365	.9	10	353	89	335	357	.8	16	330	88	280	259	.8	23
						80 HW							80 HW							328	86	305	367	.9	17	311	86	260	278	.9	25
397	86	495	397	1.0	0	40 HW	357	84	395	383	1.0	5	40 HW	328	83	323	377	1.0	11	303	83	275	376	1.0	18	294	85	247	395	1.0	27
375	83	455	415	1.1	0	0	331	81	380	416	1.1	5	0	313	81	300	400	1.1	12	278	80	245	385	1.1	19	284	84	240	420	1.1	28
						40 TW							40 TW							256	78	225	398	1.3	21	276	83	225	450	1.2	29
						80 TW							80 TW																		
						120 TW							120 TW																		

Figure A-7 (Sheet 1 of 2 Sheets) — Flight Operation Instruction Chart

RESTRICTED
AN 01-75FJC-1

HIGH ALTITUDE

AIRPLANE MOD. T-33A   **ENG.** J33-A-23, J33-A-35   **CHART WT. LIMITS:** 11,800 TO 8200 LB.   **EXT. LOAD:** NONE   **NO. OF ENGINES OPERATING:** ONE

IF YOU ARE AT 25,000'			FUEL U.S. GAL.	IF YOU ARE AT 30,000'			IF YOU ARE AT 35,000'			IF YOU ARE AT 40,000'			FUEL U.S. GAL.	IF YOU ARE AT 45,000'		
RANGE IN AIRMILES				RANGE IN AIRMILES			RANGE IN AIRMILES			RANGE IN AIRMILES				RANGE IN AIRMILES		
BY CRUISING AT 25,000	OPT ALT 1000 FT	BY CRUISING AT OPT ALT		BY CRUISING AT 30,000	OPT ALT 1000 FT	BY CRUISING AT OPT ALT	BY CRUISING AT 35,000	OPT ALT 1000 FT	BY CRUISING AT OPT ALT	BY CRUISING AT 40,000	OPT ALT 1000 FT	BY CRUISING AT OPT ALT		BY CRUISING AT 45,000	OPT ALT 1000 FT	BY CRUISING AT OPT ALT
675	40		350	770	40	940	895	40	960	975	40	—	350			
620	40		320	710	40	860	820	40	880	890	40	—	320			
545	40		280	620	40	750	720	40	775	780	40	—	280			
470	40		240	535	40	640	620	40	665	675	40	—	240			
395	40		200	450	40	535	520	40	560	570	40	—	200			
320	40		160	360	40	425	420	40	450	450	40	—	160			
240	40		120	275	40	320	320	40	340	350	40	—	120			
165	40		80	190	40	210	220	40	235	240	40	—	80			
90	—		40	105	—	—	120	—	—	140	—	—	40			

(RANGE FIGURES INCLUDE ALLOWANCE FOR PRESCRIBED CLIMB AND DESCENT TO SEA LEVEL)

	CRUISING AT 25,000						EFFEC. TIVE WIND MPH		CRUISING AT 30,000						CRUISING AT 35,000						CRUISING AT 40,000						EFFEC. TIVE WIND MPH		CRUISING AT 45,000					
	APPROXIMATE								APPROXIMATE						APPROXIMATE						APPROXIMATE								APPROXIMATE					
CAS	% RPM	GPH	G.S.	RANGE FACTOR	Let Down Dist.				CAS	% RPM	GPH	G.S.	RANGE FACTOR	Let Down Dist.	CAS	% RPM	GPH	G.S.	RANGE FACTOR	Let Down Dist.	CAS	% RPM	GPH	G.S.	RANGE FACTOR	Let Down Dist.			CAS	% RPM	GPH	G.S.	RANGE FACTOR	Let Down Dist.
314	90	250	334	.7	31		120 HW		290	91	220	340	.7	60	270	91	190	343	.7	60	252	95	185	364	.7	85		120 HW						
304	89	240	361	.8	33		80 HW		284	89	211	370	.8	64	266	90	185	378	.8	64	252	95	185	404	.8	89		80 HW						
297	88	230	392	.9	35		40 HW		284	89	211	410	.9	67	266	90	185	418	.9	67	247	94	175	435	.9	94		40 HW						
290	87	225	422	1.0	37		0		278	88	203	440	1.0	70	266	90	185	458	1.0	70	247	94	175	475	1.0	98		0						
290	87	225	462	1.1	39		40 TW		278	88	203	480	1.1	74	266	90	185	498	1.1	74	247	94	175	515	1.1	103		40 TW						
290	87	225	502	1.2	41		80 TW		270	86	185	513	1.2	77	257	89	175	524	1.2	77	247	94	175	555	1.2	107		80 TW						
281	86	215	528	1.3	43		120 TW		270	86	185	553	1.3	80	257	89	175	564	1.3	80	244	93	170	589	1.3	111		120 TW						

SPECIAL NOTES

1. Climb at 100% RPM.
2. Multiply statute units by .87 to obtain nautical units.
3. Range and fuel consumption are 5% conservative to allow for variations in service aircraft and operating techniques.
4. Read lower half of chart opposite effective wind only.
5. Make additional allowances for landing, navigational errors, combat, formation flight, etc. as required.
6. When using JP-3 fuel, for normal operation multiply the range by .96. When using gasoline (AN-F-48), for normal operation multiply the range by .88.

EXAMPLE

If you are at 10,000 feet with 280 gallons of available fuel, you can fly 335 statute airmiles by holding 328 MPH CAS. However, you can fly 650 statute airmiles by immediately climbing to 40,000 feet using 100% RPM. At 40,000 feet cruise at 247 MPH CAS and start letdown 98 statute airmiles from destination. With an 80 MPH headwind the range of 40,000 feet will be 0.8 × 650 or 520 statute airmiles. Cruise at 252 MPH CAS with this wind and start letdown 89 statute airmiles from destination.

BASED ON MIL-F-5516 (AN-F-32) GRADE JP-1 FUEL

LEGEND

EFFECTIVE WIND — HW, HEADWIND; TW, TAILWIND
RANGE FACTOR = GROUND DISTANCE (Effective Wind) / RANGE IN AIRMILES (Zero Wind)
G.S. — GROUND SPEED IN STATUTE MILES PER HOUR
CAS — CALIBRATED AIRSPEED IN STATUTE MILES PER HOUR
GPH — GALLONS PER HOUR
RANGE — STATUTE MILES

DATA AS OF: 1-1-49 BASED ON: FLIGHT TEST (With J33-A-23 Engines)   RED FIGURES HAVE NOT BEEN FLIGHT CHECKED

Figure A-7 (Sheet 2 of 2 Sheets) — Flight Operation Instruction Chart

Appendix I

RESTRICTED
AN 01-75FJC-1

AIRPLANE MODEL(S) T-33A

ENGINE(S) J33-A-23, J33-A-35

FLIGHT OPERATION INSTRUCTION CHART

CHART WEIGHT LIMITS: 14,300 TO 11,800 POUNDS

EXTERNAL LOAD ITEMS
2 — 165 GALLON EXTERNAL TIP TANKS
DROPPED WHEN EMPTY
NUMBER OF ENGINES OPERATING: ONE

INSTRUCTIONS FOR USING CHART. (A) IN FLIGHT — Select figure in fuel volume equal to or less than fuel available for cruise (fuel on board minus allowances for reserve, combat, navigational errors, formation flight, etc.). Move horizontally right or left to section according to present altitude and read total range available (no wind) by cruising at that altitude or by climbing to another altitude of maximum range. For a flight at initial altitude, succeeding instructions are given directly below. For a flight at higher altitude, climb immediately to desired altitude and read cruising instructions in appropriate cruising altitude section. **(B) FLIGHT PLANNING** — From initial fuel on board subtract fuel required for take-off and climb to desired cruising altitude and all other necessary allowances. Then use chart as for IN FLIGHT above, adding initial climb distances to range values.

NOTES: Ranges shown at optimum altitudes are maximum. In order to obtain maximum range on flights requiring more than one chart (due to external configuration or gross weight change), it is necessary to observe the optimum cruising altitude on each chart; i.e., when changing charts a climb may be required to obtain a maximum range. All range values include allowances for descent distance and fuel. Climb distance and fuel are included where climbs are indicated.

DATA BELOW CONTAIN NO FUEL RESERVE FOR LANDING

LOW ALTITUDE

IF YOU ARE AT S.L.				FUEL U.S. GAL.	IF YOU ARE AT 5000'				IF YOU ARE AT 10,000'				IF YOU ARE AT 15,000'				FUEL U.S. GAL.	IF YOU ARE AT 20,000'			
RANGE IN AIRMILES		RANGE FACTOR	Let Down Dist.		RANGE IN AIRMILES		RANGE FACTOR	Let Down Dist.	RANGE IN AIRMILES		RANGE FACTOR	Let Down Dist.	RANGE IN AIRMILES		RANGE FACTOR	Let Down Dist.		RANGE IN AIRMILES		RANGE FACTOR	Let Down Dist.
BY CRUISING AT S.L.	OPT ALT 1000 FT				BY CRUISING AT 5000'	OPT ALT 1000 FT			BY CRUISING AT 10,000'	OPT ALT 1000 FT			BY CRUISING AT 15,000'	OPT ALT 1000 FT				BY CRUISING AT 20,000'	OPT ALT 1000 FT		
550	40			700	660	40			775	40			940	40			700	1080	40		
510	40			650	615	40			725	40			875	40			650	1010	40		
470	40			600	570	40			670	40			810	40			600	940	40		
435	40			550	525	40			620	40			750	40			550	870	40		
395	40			500	480	40			570	40			685	40			500	795	40		
360	40			450	435	40			520	40			620	40			450	720	40		
320	40			400	390	40			470	40			560	40			400	650	40		
285	40			350	345	40			420	40			495	40			350	575	40		

(RANGE FIGURES INCLUDE ALLOWANCE FOR PRESCRIBED CLIMB AND DESCENT TO SEA LEVEL)

CRUISING AT S.L.					EFFECTIVE WIND MPH	CRUISING AT 5000'					CRUISING AT 10,000'					CRUISING AT 15,000'					EFFECTIVE WIND MPH	CRUISING AT 20,000'				
		APPROXIMATE						APPROXIMATE					APPROXIMATE					APPROXIMATE						APPROXIMATE		
CAS	% RPM	GPH	G.S.	RANGE FACTOR		CAS	% RPM	GPH	G.S.	RANGE FACTOR	CAS	% RPM	GPH	G.S.	RANGE FACTOR	CAS	% RPM	GPH	G.S.	RANGE FACTOR		CAS	% RPM	GPH	G.S.	RANGE FACTOR
					120 HW																120 HW					
					80 HW																80 HW					
383	87	520	347	.9	40 HW	362	87	430	348	.9	340	88	385	350	.9	316	87	320	312	.8	40 HW	310	87	290	335	.8
374	86	500	376	1.0	0	353	86	425	382	1.0	330	87	375	375	1.0	308	86	305	343	.9	0	300	88	280	365	.9
365	85	480	404	1.1	40 TW	344	85	415	408	1.1	320	86	360	407	1.1	298	85	295	371	1.0	40 TW	290	87	270	390	1.0
					80 TW											291	84	285	402	1.1	80 TW	280	86	260	415	1.1
					120 TW											282	83	275	432	1.2	120 TW	270	85	250	440	1.2

Let Down Dist: 0, 0, 0, 4, 4, 4, 8, 9, 10, 14, 15, 16, 17, 18, 21, 23, 24, 25, 26

Figure A-7A (Sheet 1 of 2 Sheets) — Flight Operation Instruction Chart

48   RESTRICTED   Revised 15 November 1950

Appendix I

HIGH ALTITUDE

AIRPLANE MOD. T33A ENG. J33-A-23, J33-A-35 CHART WT. LIMITS: 14,300 TO 11,800 LB. EXT. LOAD: 2 x 165 GAL. EXTERNAL TIP TANKS NO. OF ENGINES OPERATING: ONE DROPPED WHEN EMPTY

IF YOU ARE AT 25,000'			FUEL U.S. GAL.	IF YOU ARE AT 30,000'			IF YOU ARE AT 35,000'			IF YOU ARE AT 40,000'			FUEL U.S. GAL.	IF YOU ARE AT 45,000'		
RANGE IN AIRMILES				RANGE IN AIRMILES			RANGE IN AIRMILES			RANGE IN AIRMILES				RANGE IN AIRMILES		
BY CRUISING AT 25,000'	OPT.ALT 1000 FT.	BY CRUISING AT OPT.ALT.		BY CRUISING AT 30,000'	OPT.ALT 1000 FT.	BY CRUISING AT OPT.ALT.	BY CRUISING AT 35,000'	OPT.ALT 1000 FT.	BY CRUISING AT OPT.ALT.	BY CRUISING AT 40,000'	OPT.ALT 1000 FT.	BY CRUISING AT OPT.ALT.		BY CRUISING AT 45,000'	OPT.ALT 1000 FT.	BY CRUISING AT OPT.ALT.
1250	40		700	1430	40		1650	40		1830	—	—	700			
1170	40		650	1335	40		1540	40		1710	—	—	650			
1090	40		600	1240	40		1435	40		1585	—	—	600			
1005	40		550	1145	40		1325	40		1465	—	—	550			
925	40		500	1050	40		1215	40		1345	—	—	500			
845	40		450	955	40		1110	40		1220	—	—	450			
760	40		400	860	40		1000	40		1095	—	—	400			
675	40		350	770	40		895	40		975	—	—	350			

(RANGE FIGURES INCLUDE ALLOWANCE FOR PRESCRIBED CLIMB AND DESCENT TO SEA LEVEL)

CRUISING AT 25,000'						EFFEC-TIVE WIND MPH	CRUISING AT 30,000'						CRUISING AT 35,000'						CRUISING AT 40,000'						EFFEC-TIVE WIND MPH
		APPROXIMATE							APPROXIMATE						APPROXIMATE						APPROXIMATE				
CAS	% RPM	GPH	G.S.	RANGE FACTOR	Let Down Dist.		CAS	% RPM	GPH	G.S.	RANGE FACTOR	Let Down Dist.	CAS	% RPM	GPH	G.S.	RANGE FACTOR	Let Down Dist.	CAS	% RPM	GPH	G.S.	RANGE FACTOR	Let Down Dist.	
301	91	275	316	.7	30	120 HW	280	92	240	320	.7	38	260	93	215	325	.7	55	240	96	195	343	.7	73	120 HW
292	90	260	345	.8	31	80 HW	270	91	230	350	.8	41	253	92	205	356	.8	57	240	96	195	383	.8	77	80 HW
292	90	260	385	.9	32	40 HW	270	91	230	390	.9	44	253	92	205	396	.9	59	240	96	195	423	.9	81	40 HW
283	89	250	412	1.0	34	0	260	90	220	416	1.0	46	244	91	195	421	1.0	62	234	95	185	452	1.0	85	0
273	88	240	438	1.1	36	40 TW	255	89	210	445	1.1	48	244	91	195	461	1.1	65	234	95	185	492	1.1	89	40 TW
273	88	240	478	1.2	38	80 TW	255	89	210	485	1.2	51	235	90	185	488	1.2	69	228	94	175	520	1.2	93	80 TW
263	87	225	504	1.3	40	120 TW	250	88	200	515	1.3	53	235	90	185	528	1.3	72	228	94	175	560	1.3	98	120 TW

SPECIAL NOTES

1 Climb at 100% RPM.
2 Multiply statute units by .87 to obtain nautical units.
3 Range and fuel consumption are 5% conservative to allow for variations in service aircraft and operating techniques.
4 Read lower half of chart opposite effective wind only.
5 Make additional allowances for landing, navigational errors, combat, formation flight, etc. as required.
6 When using JP-3 fuel, for normal operation multiply the range by .90. When using gasoline (AN-F-48), for normal operation multiply the range by .88.

EXAMPLE

If you are at 10,000 feet with 600 gallons of available fuel, you can fly 670 statute airmiles by holding 330 MPH CAS. However, you can fly 1435 statute airmiles by immediately climbing to 40,000 feet using 100% RPM. At 40,000 feet cruise of 234 MPH CAS and start letdown 85 statute airmiles from destination. With an 80 MPH headwind the range at 40,000 feet will be 1435 x .8 or 1148 airmiles. Cruise at 240 MPH CAS with this wind and start letdown 77 statute airmiles from destination.

BASED ON MIL-F-5516 (AN-F-32) GRADE JP-1 FUEL

DATA AS OF: 1-1-49 BASED ON: FLIGHT TEST (With J33-A-23 Engines) RED FIGURES HAVE NOT BEEN FLIGHT CHECKED

LEGEND

EFFECTIVE WIND — HW, HEADWIND; TW, TAILWIND
GROUND DISTANCE (Effective Wind)
RANGE FACTOR = RANGE IN AIRMILES (Zero Wind)
G.S. — GROUND SPEED IN STATUTE MILES PER HOUR
CAS — CALIBRATED AIRSPEED IN STATUTE MILES PER HOUR
GPH — GALLONS PER HOUR
RANGE — STATUTE MILES

Figure A-7A (Sheet 2 of 2 Sheets) — Flight Operation Instruction Chart

FLIGHT OPERATION INSTRUCTION CHART

AIRPLANE MODEL(S): T-33A
ENGINE(S): J33-A-23, J33-A-35

EXTERNAL LOAD ITEMS: 2 – 165 GALLON EXTERNAL TIP TANKS CARRIED ALL THE WAY
NUMBER OF ENGINES OPERATING: ONE

CHART WEIGHT LIMITS: 14,300 TO 8500 POUNDS

INSTRUCTIONS FOR USING CHART: (A) IN FLIGHT — Select figure in fuel column equal to or less than fuel available for cruise (fuel on board minus allowance for reserve, combat, navigational errors, formation flight, etc.). Move horizontally right or left to section according to present altitude and read total range available (no wind) by cruising at that altitude or by climbing to another altitude of maximum range. For a flight at higher altitude, operating instructions are given directly below. For a flight at higher altitude, climb immediately to desired altitude and read cruising instructions in appropriate cruising altitude section. (B) FLIGHT PLANNING — From initial fuel on board subtract fuel required for take-off and climb to desired cruising altitude and all other necessary allowances. Then use chart as for IN FLIGHT above, adding initial climb distances to range values.

NOTES: Ranges shown at optimum altitudes are maximum. In order to obtain maximum range on flights requiring more than one chart (due to external configuration or gross weight change), it is necessary to observe the optimum cruising altitude on each chart, i.e., when changing charts a climb may be required to obtain a maximum range. All range values include allowances for climb distance and fuel. Climb distance and fuel are included where climbs are indicated.

DATA BELOW CONTAIN NO FUEL RESERVE FOR LANDING

LOW ALTITUDE

IF YOU ARE AT S.L.			FUEL U.S. GAL	IF YOU ARE AT 5000'			IF YOU ARE AT 10,000'			IF YOU ARE AT 15,000'			FUEL U.S. GAL	IF YOU ARE AT 20,000'		
RANGE IN AIRMILES				RANGE IN AIRMILES			RANGE IN AIRMILES			RANGE IN AIRMILES				RANGE IN AIRMILES		
BY CRUISING AT S.L.	OPT.ALT. 1000 FT.	BY CRUISING AT OPT.ALT.		BY CRUISING AT 5000'	OPT.ALT. 1000 FT.	BY CRUISING AT OPT.ALT.	BY CRUISING AT 10,000'	OPT.ALT. 1000 FT.	BY CRUISING AT OPT.ALT.	BY CRUISING AT 15,000'	OPT.ALT. 1000 FT.	BY CRUISING AT OPT.ALT.		BY CRUISING AT 20,000'	OPT.ALT. 1000 FT.	BY CRUISING AT OPT.ALT.
525	40	1535	700	625	40	1560	715	40	1595	890	40	1620	700	1030	40	1640
450	40	1290	600	535	40	1320	615	40	1355	760	40	1385	600	880	40	1400
375	40	1045	500	450	40	1080	510	40	1110	635	40	1140	500	735	40	1155
335	40	925	450	405	40	955	460	40	985	575	40	1015	450	665	40	1035
300	40	800	400	360	40	835	410	40	865	510	40	895	400	590	40	915
260	40	680	350	315	40	715	360	40	745	445	40	770	350	520	40	790
225	40	560	300	270	40	595	310	40	620	385	40	650	300	445	40	670
185	35	415	250	225	40	470	260	40	495	320	40	525	250	375	40	550
150	35	300	200	180	35	330	210	35	360	260	35	380	200	300	35	405
110	25	190	150	135	25	215	160	35	250	195	35	275	150	230	35	300
75	25	80	100	90	25	135	110	35	150	130	35	170	100	155	35	195

CRUISING AT S.L.					CRUISING AT 5000'					CRUISING AT 10,000'					CRUISING AT 15,000'					EFFEC-TIVE WIND MPH	CRUISING AT 20,000'				
APPROXIMATE			RANGE FACTOR	Let Down Dist.	APPROXIMATE			RANGE FACTOR	Let Down Dist.	APPROXIMATE			RANGE FACTOR	Let Down Dist.	APPROXIMATE			RANGE FACTOR	Let Down Dist.		APPROXIMATE			RANGE FACTOR	Let Down Dist.
CAS	% RPM	GPH G.S.			CAS	% RPM	GPH G.S.			CAS	% RPM	GPH G.S.			CAS	% RPM	GPH G.S.				CAS	% RPM	GPH G.S.		
																				120 HW					
383	87	520 347	.9	0	362	87	430 348	.9	4	340	88	365 350	.9	8	316	87	320 312	.8	14	80 HW	310	89	290 335	.8	21
374	86	500 376	1.0	0	353	86	425 382	1.0	4	330	87	375 375	1.0	9	308	86	305 343	.9	15	40 HW	300	88	280 365	.9	23
365	85	480 404	1.1	0	344	85	415 408	1.1	4	320	86	360 407	1.1	10	298	85	295 371	1.0	16	0	290	87	270 390	1.0	24
															291	84	285 402	1.1	17	40 TW	280	86	260 415	1.1	25
															282	83	275 432	1.2	18	80 TW	270	85	250 440	1.2	26
																				120 TW					

(RANGE FIGURES INCLUDE ALLOWANCE FOR PRESCRIBED CLIMB AND DESCENT TO SEA LEVEL)

Figure A-7B (Sheet 1 of 2 Sheets) — Flight Operation Instruction Chart

RESTRICTED
AN 01-75FJC-1

Appendix I

HIGH ALTITUDE

AIRPLANE MOD. T-33A   ENG. J33-A-23, J33-A-35   CHART WT. LIMITS: 14,300 TO 8500 LB.   EXT. LOAD: 2 — 165 GAL. TIP TANKS CARRIED ALL THE WAY   NO. OF ENGINES OPERATING: ONE

IF YOU ARE AT 25,000'			FUEL U.S. GAL	IF YOU ARE AT 30,000'			IF YOU ARE AT 35,000'			IF YOU ARE AT 40,000'			FUEL U.S. GAL	IF YOU ARE AT 45,000'		
RANGE IN AIRMILES				RANGE IN AIRMILES			RANGE IN AIRMILES			RANGE IN AIRMILES				RANGE IN AIRMILES		
BY CRUISING AT 25,000'	OPT ALT 1000 FT.	BY CRUISING AT OPT ALT.		BY CRUISING AT 30,000'	OPT ALT 1000 FT.	BY CRUISING AT OPT ALT.	BY CRUISING AT 35,000'	OPT ALT 1000 FT.	BY CRUISING AT OPT ALT.	BY CRUISING AT 40,000'	OPT ALT 1000 FT.	BY CRUISING AT OPT ALT.		BY CRUISING AT 45,000'	OPT ALT 1000 FT.	BY CRUISING AT OPT ALT.
1180	40	1680	700	1160	40	1470	1305	40	1485	1495			700			
1015	40	1440	600										600			
845	40	1220	500	970	40	1220	1090	40	1245	1250	—	—	500			
765	40	1080	450	875	40	1100	985	40	1125	1135	—	—	450			
680	40	955	400	760	40	975	875	40	1000	1015	—	—	400			
				(RANGE FIGURES INCLUDE ALLOWANCE FOR PRESCRIBED CLIMB AND DESCENT TO SEA LEVEL)												
595	40	835	350	685	40	850	770	40	875	890	—	—	350			
515	40	715	300	590	40	730	665	40	755	770	—	—	300			
430	40	595	250	495	40	605	555	40	630	650	—	—	250			
350	40	460	200	400	40	485	450	40	505	530	—	—	200			
265	40	340	150	305	40	365	340	40	385	405	—	—	150			
180	35	215	100	210	40	—	235	40	255	285	—	—	100			

CRUISING AT 25,000							EFFEC- TIVE WIND MPH	CRUISING AT 30,000'							CRUISING AT 35,000'							CRUISING AT 40,000'							EFFEC- TIVE WIND MPH	CRUISING AT 45,000'						
APPROXIMATE								APPROXIMATE							APPROXIMATE							APPROXIMATE								APPROXIMATE						
CAS	% RPM	GPH	G.S.	RANGE FACTOR	Let Down Dist.			CAS	% RPM	GPH	G.S.	RANGE FACTOR	Let Down Dist.		CAS	% RPM	GPH	G.S.	RANGE FACTOR	Let Down Dist.		CAS	% RPM	GPH	G.S.	RANGE FACTOR	Let Down Dist.			CAS	% RPM	GPH	G.S.	RANGE FACTOR	Let Down Dist.
301	91	275	316	.7	30		120 HW	280	92	240	320	.7	38		260	93	215	325	.7	55		240	96	195	343	.7	73		120 HW						
292	90	260	345	.8	31		80 HW	270	91	230	350	.8	41		253	92	205	356	.8	57		240	96	195	383	.8	77		80 HW						
292	90	260	385	.9	32		40 HW	270	91	230	390	.9	44		253	92	205	396	.9	59		240	96	195	423	.9	81		40 HW						
283	89	250	412	1.0	34		0	260	90	220	416	1.0	46		244	91	195	421	1.0	62		234	95	185	452	1.0	85		0						
273	88	240	438	1.1	36		40 TW	255	89	210	445	1.1	48		244	91	195	461	1.1	65		234	95	185	492	1.1	89		40 TW						
273	88	240	478	1.2	38		80 TW	255	89	210	485	1.2	51		235	90	185	488	1.2	69		228	94	175	520	1.2	93		80 TW						
263	87	225	504	1.3	40		120 TW	250	88	200	515	1.3	53		235	90	185	528	1.3	72		220	94	175	560	1.3	98		120 TW						

SPECIAL NOTES

1 – Climb at 100% RPM.
2 – Multiply statute units by .87 to obtain nautical units.
3 – Range and fuel consumption are 5% conservative to allow for variations in service aircraft and operating techniques.
4 – Read lower half of chart opposite effective wind only.
5 – Make additional allowances for landing, navigational errors, combat, formation flight, etc. as required.
6 – When using JP-3 fuel, for normal operation multiply the range by .96. When using gasoline (AN-F-48), for normal operation multiply the range by .88.

EXAMPLE

If you are at 10,000 feet with 600 gallons of available fuel, you can fly 615 statute airmiles by holding 330 MPH CAS. However, you can fly 1355 statute airmiles by immediately climbing to 40,000 feet using 100% RPM. At 40,000 feet cruise at 234 MPH CAS and start letdown 85 statute airmiles from destination. With an 80 MPH headwind the range at 40,000 feet will be 0.8 x 1355 or 1084 statute airmiles. Cruise at 240 MPH CAS with this wind and start letdown 77 statute airmiles from destination.

BASED ON MIL-F-5516 (AN-F-32) GRADE JP-1 FUEL

LEGEND

EFFECTIVE WIND – HW, HEADWIND; TW, TAILWIND
EFFECTIVE WIND – GROUND DISTANCE (Effective Wind)
RANGE FACTOR = RANGE IN AIRMILES (Zero Wind)
G.S. – GROUND SPEED IN STATUTE MILES PER HOUR
CAS – CALIBRATED AIRSPEED IN STATUTE MILES PER HOUR
GPH – GALLONS PER HOUR
RANGE – STATUTE MILES

DATA AS OF: 1-1-49   BASED ON: FLIGHT TEST (With J33-A-23 Engines)   RED FIGURES HAVE NOT BEEN FLIGHT CHECKED

Figure A-7B (Sheet 2 of 2 Sheets) — Flight Operation Instruction Chart

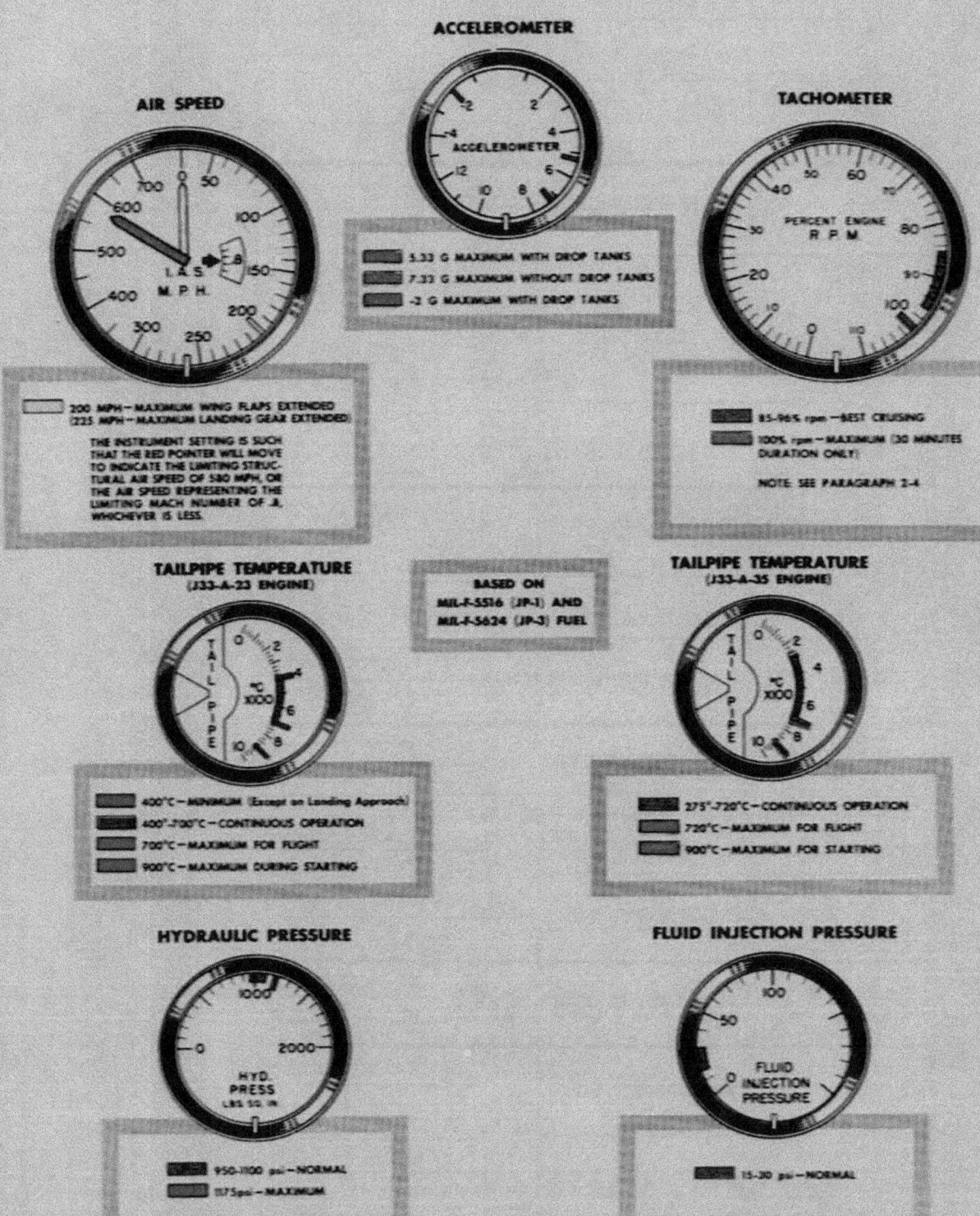

Figure A-8 (Sheet 1 of 2 Sheets) — Instrument Markings Diagram

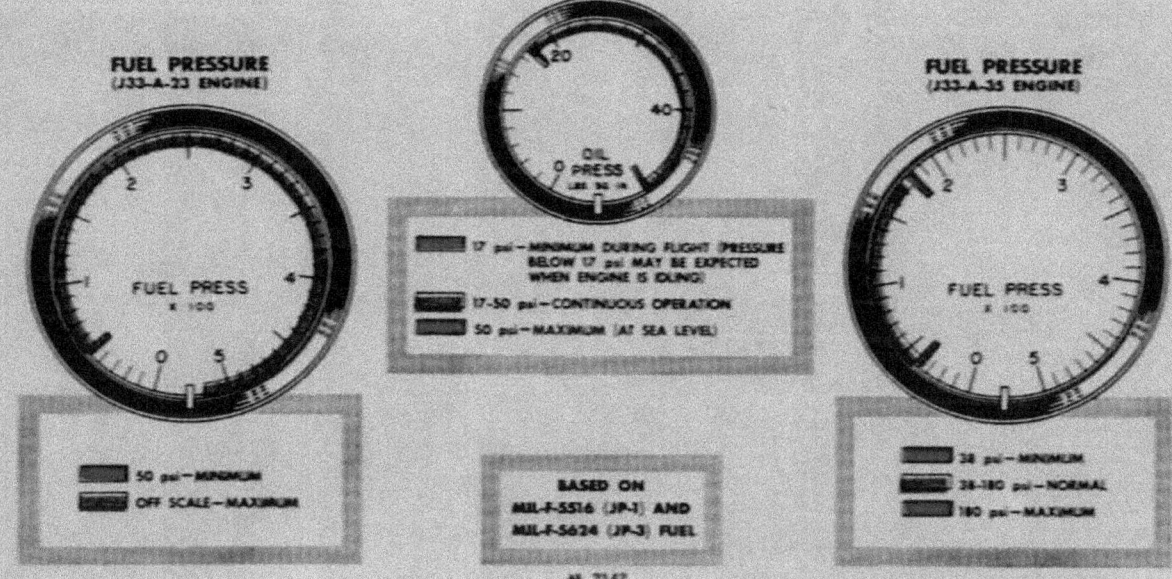

Figure A-8 (Sheet 2 of 2 Sheets) — Instrument Markings Diagram

THE FLEET TYPE SUBMARINE
The definitive technical guide – profiling the submarines of World War II
RESTRICTED
NOW ALSO AVAILABLE FROM WWW.PERISCOPEFILM.COM:
THE DEFINITIVE GUIDE TO THE SUBMARINES THAT WON WWII
Originally published as NAVPERS 16160
June 1946
WWW.PERISCOPEFILM.COM

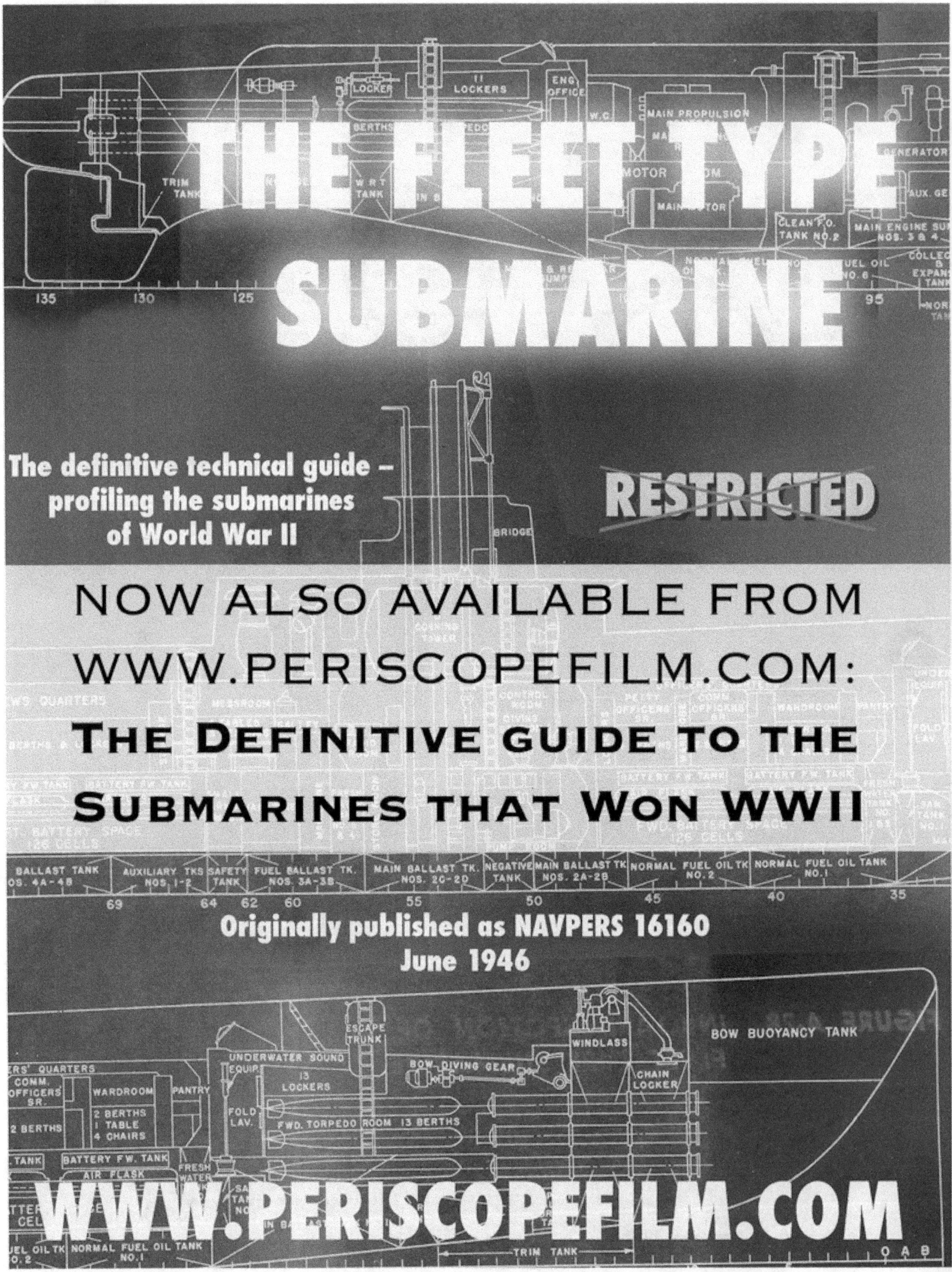

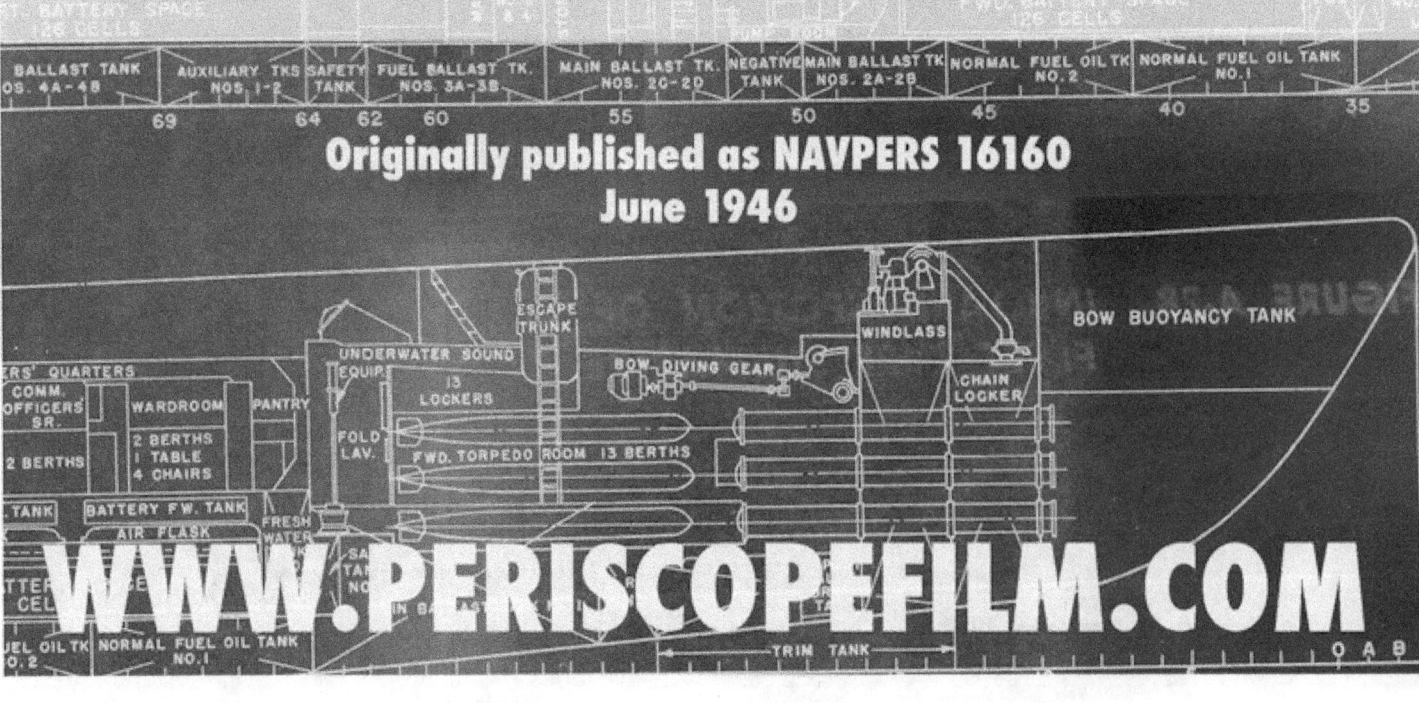

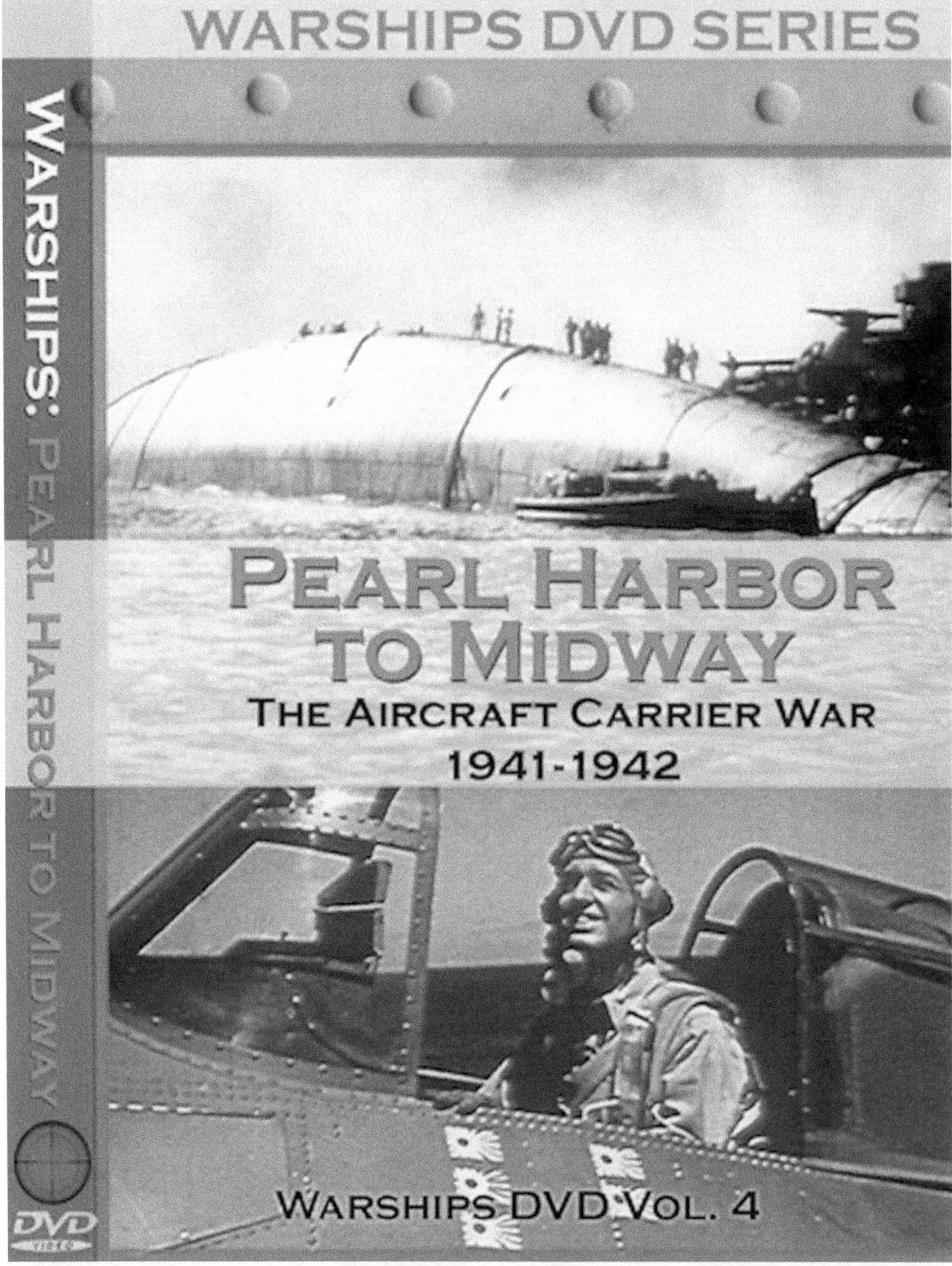

WARSHIPS DVD SERIES

AIRCRAFT CARRIER MISHAPS
SAFETY AND TRAINING FILMS

-PeriscopeFilm.com-

Now Available on DVD!

Epic Battles of WWII

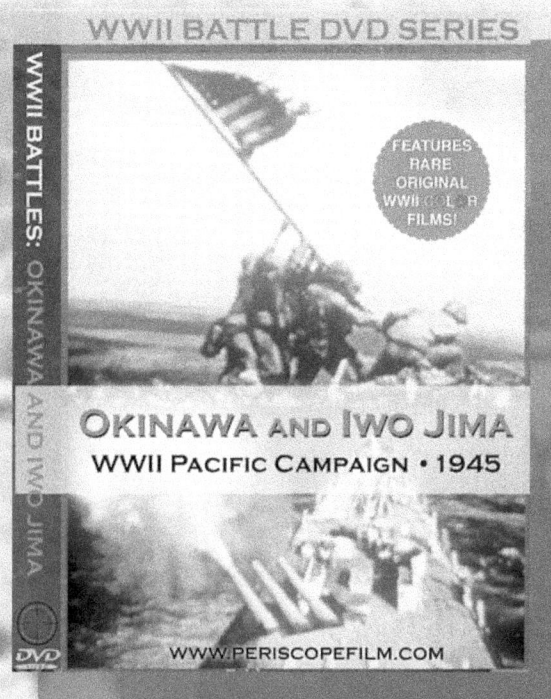

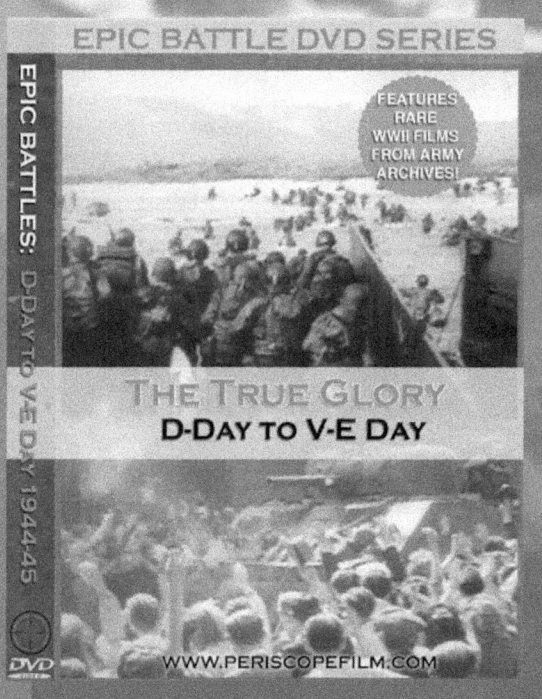

Now Available on DVD!

Warships DVD Series

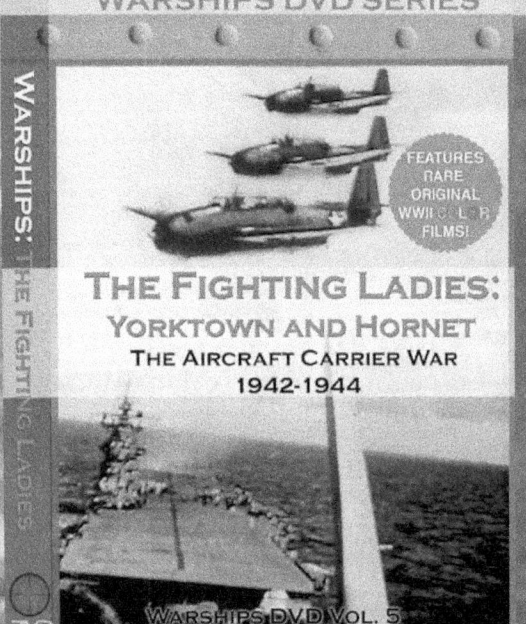

Now Available!

Aircraft At War DVD Series

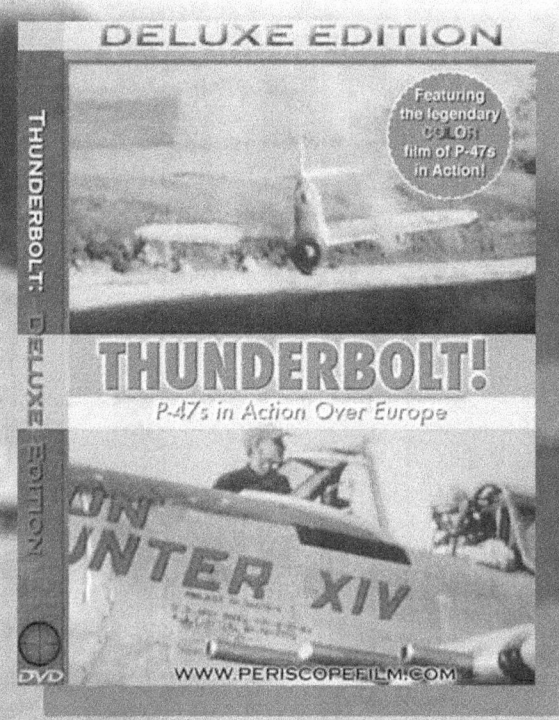

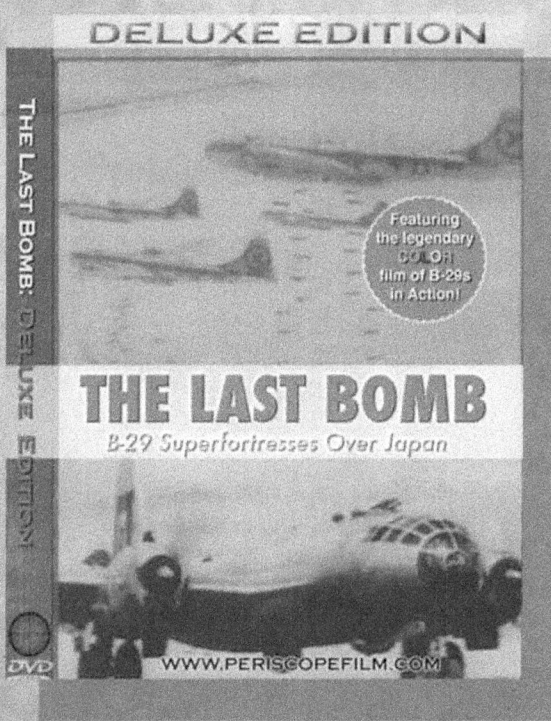

Now Available!

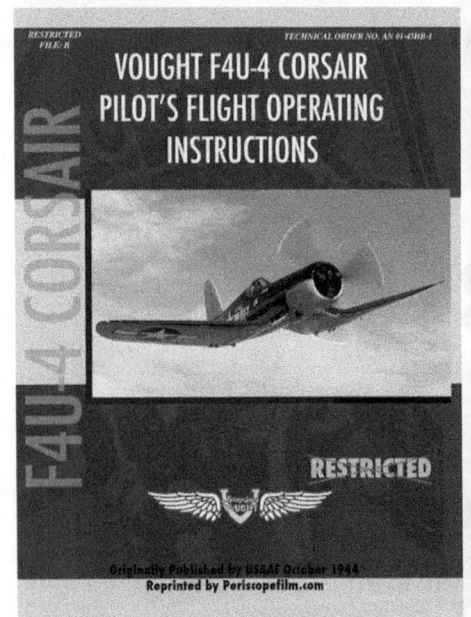
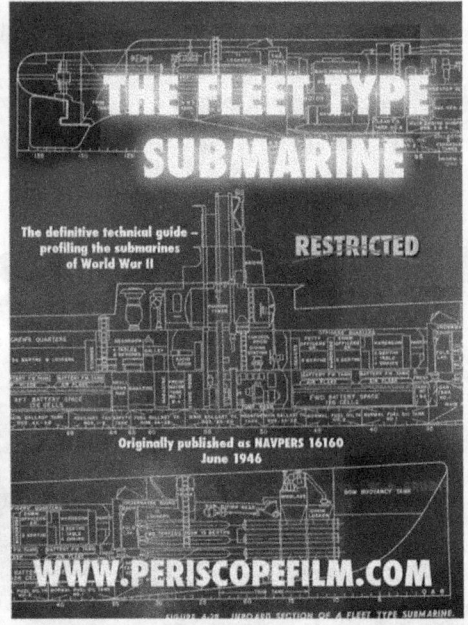
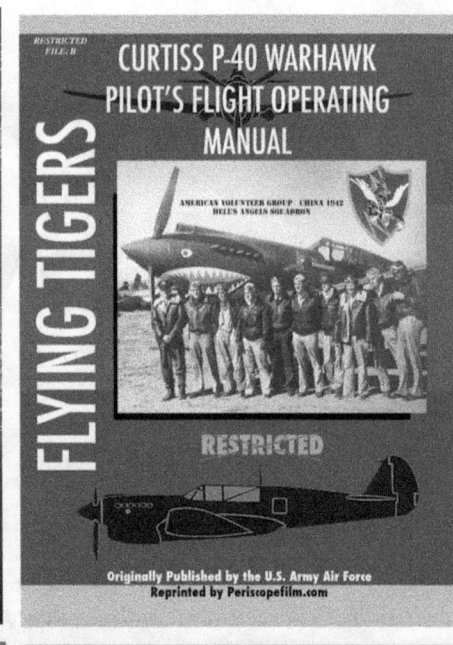

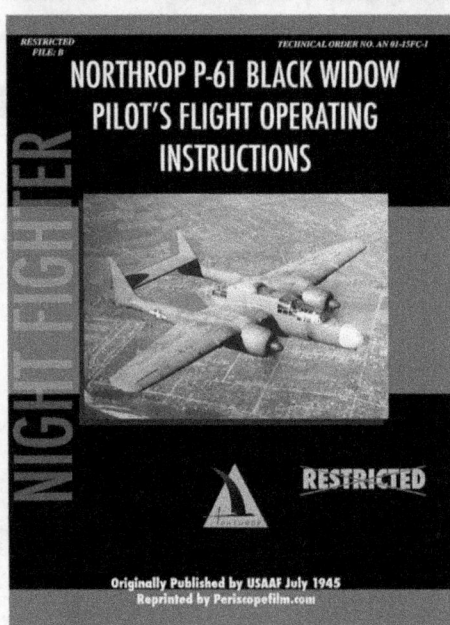

ALSO NOW AVAILABLE
FROM PERISCOPEFILM.COM

ISBN #978-1-935327-51-6 1-935327-51-8
©2008 Periscope Film LLC
All Rights Reserved
www.PeriscopeFilm.com

www.ingramcontent.com/pod-product-compliance
Lightning Source LLC
Chambersburg PA
CBHW080524110426
42742CB00017B/3228